The Strategic Perspective and Long-Term Socioeconomic Strategies for Israel

Key Methods with an Application to Aging

Steven W. Popper, Howard J. Shatz, Shmuel Abramzon, Claude Berrebi, Shira Efron

Prepared for the Prime Minister's Office and Ministry of Finance, Government of Israel

For more information on this publication, visit www.rand.org/t/RR488

Library of Congress Cataloging-in-Publication Data is available for this publication.
ISBN: 978-0-8330-9073-7

Published by the RAND Corporation, Santa Monica, Calif.
© Copyright 2015 RAND Corporation
RAND® is a registered trademark.

Cover image: Getty Images (Dan Porges)

Support RAND
Make a tax-deductible charitable contribution at
www.rand.org/giving/contribute

www.rand.org

Preface

Many of Israel's socioeconomic achievements have been dramatic. A young country that has absorbed massive influxes of immigrants and responded to constant security threats, Israel successfully developed a vibrant, open, and technologically advanced economy characterized by high rates of economic growth. These strengths were well manifested during the recent global downturn in which Israel's economy outperformed that of almost all other developed countries.

However, Israel faces substantial economic and social challenges, some of which are well known to policymakers and have recently fueled a wide wave of social unrest. These challenges include differential ability within the population to participate in and benefit from the growth in the economy, rising costs of living, and questions on the part of the public about the government's ability to address those challenges. Faced with major existential challenges throughout Israel's history, the government has not routinely developed strategic responses to problems that demand longer-term, coordinated policy action.

Israel's National Economic Council (NEC) and Ministry of Finance (MOF) asked the RAND Corporation, a global policy research institution, to assist a select task force of leading government officials in developing means by which Israel could enhance its capacity for applying strategic thinking to the development of government policy, particularly in the socioeconomic sphere. They asked us to provide a range of inputs on such matters as the nature of strategic planning in government; issues of measurement and assessment; methods and techniques for strategic assessment and strategic plan development; appropriate institutions, processes, and data sources for enhancing capacities for strategic assessment; and comparisons with the practice in other countries. Another publication (Shatz et al., 2015) discusses institutions and process.

This report expands on three further RAND inputs to the task force: the role for a strategic perspective as part of the policy planning process in Israel; the challenge of defining socioeconomic goals compatible with an underlying vision; and how to apply a strategic perspective, using as an example an issue (the aging of Israel's population) that crosses short and longer time horizons, as well as the responsibilities of many ministries. Our purpose is to provide a summary document for policymakers and staff that captures some of the mechanics of bringing a strategic perspective to socioeconomic strategy and provides an illustrative application.

The NEC in the Prime Minister's Office (PMO) and the MOF sponsored the research reported here. This report should be of interest to the Israeli public at large, government officials in Israel responsible for policy design, and researchers and others who have an interest in comparative governance and strategy.

RAND Labor and Population and the RAND Environment, Energy, and Economic Development Program

This research was undertaken jointly within RAND Labor and Population and the Environment, Energy, and Economic Development Program.

RAND Labor and Population has built an international reputation for conducting objective, high-quality, empirical research to support and improve policies and organizations around the world. Its work focuses on children and families, demographic behavior, education and training, labor markets, social welfare policy, immigration, international development, financial decisionmaking, and issues related to aging and retirement with a common aim of understanding how policy and social and economic forces affect individual decisionmaking and human well-being.

The RAND Environment, Energy, and Economic Development Program addresses topics relating to environmental quality and regulation, water and energy resources and systems, climate, natural hazards and disasters, and economic development, both domestically and internationally. Program research is supported by government agencies, foundations, and the private sector.

This program is part of RAND Justice, Infrastructure, and Environment, a division of the RAND Corporation dedicated to improving policy and decisionmaking in a wide range of policy domains, including civil and criminal justice, infrastructure protection and homeland security, transportation and energy policy, and environmental and natural resource policy.

Questions or comments about this report should be sent to the project leader, Steven W. Popper (Steven_Popper@rand.org). For more information about RAND Labor and Population, see www.rand.org/labor or contact the director at lpinfo@rand.org. For more information about the Environment, Energy, and Economic Development Program, see www.rand.org/energy or contact the director at eeed@rand.org.

Contents

Figures

Tables

Summary

We provided support to an Israeli government team of high-level officials charged with developing a long-term socioeconomic strategy for the state. This report highlights selected inputs we made to the government task force for the purpose of summarizing the essential mechanics and role for bringing a strategic perspective to the consideration of policy. In doing so, we provide the example of problems associated with an aging population as an illustration of how one can use a strategic perspective in an analysis of policy choices.

The Strategic Perspective and Socioeconomic Policy

Israel will profit from bringing a systemic *strategic perspective* into its policy process. The concept is integral to formal strategic planning but distinct; although the latter places emphasis on an output (a strategic plan), a strategic perspective is a process for bringing an analytical element into policy decisionmaking. A strategic perspective helps to bridge not only the gap between a short-term focus and longer-term outcomes but also that across ministerial portfolios.

We use a basic model of decisionmaking called the observe, orient, decide, and act (OODA) loop, shown in Figure S.1, as our primary structure (Angerman, 2004). Operating strategically means scanning the horizon (observing) and discovering implications (orienting). Observation and orientation thus influence decisions and actions. The circularity of this process suggests the value of having these functions be integral to the institutions involved in decisionmaking. Link-

Figure S.1
The Observe, Orient, Decide, and
Act Loop

RAND *RR488-S.1*

ing observation and orientation to decisions and actions within the context of the OODA loop largely corresponds to bringing a strategic perspective into decisionmaking.

In a companion document (Shatz et al., 2015), we assess strategic approaches to socioeconomic policy in Israel. We identify instances in which the institutions and processes of government do not adequately provide actionable observation and orientation. Enhancing these capacities would provide significant practical benefits to the state of Israel. These capacities would provide

- a system for allowing early detection of trends and opportunities
- analytical weighing of potential alternative policies and their likely consequences
- the foundation for wider discussions within the government of complex issues that are difficult to conduct on a shorter-term, crisis basis
- an ongoing process for assessing strengths, weaknesses, opportunities, and threats.

A strategic perspective requires a common frame of reference for each element of the OODA loop, grounded in what matters to decisionmakers, as shown in Figure S.2. It typically begins with a **vision**:

Figure S.2
Applying a Strategic Perspective (Observation and Orientation) Within the Framework of Long-Term Socioeconomic Decisionmaking

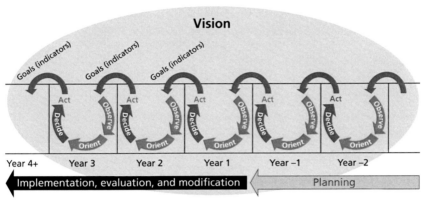

What does a desirable future state of the world look like? Translating a vision into policy requires an understanding of the challenges to achieving the vision and processes for setting specific **goals** to meet those challenges, identifying **indicators** to measure both status and progress toward goals, and designing and implementing policy measures that will contribute to achievement of goals.

The absence of meaningful vision statements at a national level can be overcome by recognizing generally agreed elements of a larger vision that are implicit in the common political discourse. The government panel asked RAND to provide a concrete example of these vision elements for illustrative purposes only. A balanced scorecard is one approach to organize goals and indicators that decisionmakers can use within an encompassing vision for the future (Kaplan and Norton, 1992, 1996). Figure S.3 provides one possible top-level framing of several vision elements by laying out a balanced scorecard for socioeconomic well-being in Israel. Not all Israelis would necessarily accept this precise formulation, but a case can be made for each account to be present in some form.

The top three elements in Figure S.3 have major influence on socioeconomic outcomes. The first element is socioeconomic well-being from the perspective of the individual and family. The second,

Figure S.3
Potential Elements of a Common Vision

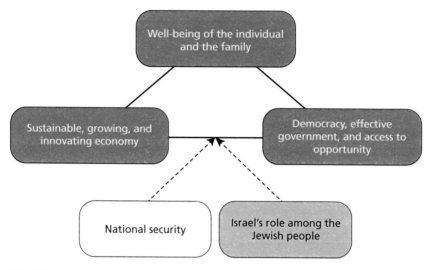

a sustainable, growing, and innovating economy, provides the balance of a national perspective to guarantee the interests of future generations. The third element emphasizes institutional structures. In Israel, issues of trust, satisfaction with government, access to opportunity, and even the fundamental character of the state routinely appear in public discourse. We have included two other elements, in lighter shades, for discussion. Israel forms the center of Jewish culture and civilization. Its policymakers make policy choices based on that role, so its omission would fail to reflect actual trade-offs. For most countries, we could omit national security from the socioeconomic scorecard. For Israel, this is a judgment call, so the figure allows others to determine what role it can play.

Each of the vision elements in Figure S.3 provides a basis for measurable goals and the indicators shown in Figure S.2 to be used in measuring their attainment. Appendix A shows, for illustrative purposes, a draft matrix of socioeconomic vision elements, goals, and sample proposed indicators. The reader should view it as an active tool for encouraging comprehensive thinking, defining what might be acceptable ranges of values for each indicator, horizon-scanning (observing) and early warning (orienting), and defining and constructing forecasts or framing scenarios.

Forming a Strategic Perspective on Population Aging

It is convenient to present the process of developing a strategic perspective as a sequence of steps. In practice, the actual transition between steps and the sequence in which they occur can become considerably more complicated. Table S.1 shows the major conceptual steps. The first six steps (rows) are sufficient for developing a strategic perspective (observe and orient in terms of the OODA loop), while the last three

Table S.1
Steps in Implementing a Strategic Perspective in Decisionmaking

OODA Stage	Step	Purpose	Example Application
Observe and orient	1. Scan and list initial questions.	Looking forward and identifying potential issues for focus	Backcasting exercise (Chapter Six); demographic trends in Israel (Chapter Seven)
	2. Gather information.	Making use of prior work and available information to create a base of current initial information	Subject-area overview (Chapter Seven); general research on aging (Chapter Seven); international experience (Chapter Seven)
	3. Understand the system.	Identifying key factors and forces and how they interact within the socioeconomic sphere	Map onto indicators (Chapter Seven); select priority indicators (Chapter Seven)
	4. Perform dynamic analysis.	Exploring trends, uncertainties, and possible trajectories	Health care–cost trend analysis (Chapter Eight); dependency ratio analysis (Chapter Eight)

Table S.1—Continued

OODA Stage	Step	Purpose	Example Application
Observe and orient, continued	5. Build scenarios.	Constructing alternative future views designed to frame dialogue and analysis	Futures wheel (passive) (Chapter Nine); futures wheel (active) (Chapter Nine); "what could go wrong?" (Chapter Nine)
	6. Identify key questions and strategic directions.	Synthesizing and integrating, which could motivate more-detailed analyses and preparation for action	Defining main questions (Chapter Ten); strategic alternatives (Chapter Ten)
Decide and act	7. Define strategic objectives.	Deriving and achieving consensus on a set of desirable outcomes	
	8. Assess alternative strategies.	Examining strategic choices systematically and building a robust course of action	
	9. Plan the implementation.	Determining requirements for moving from a strategy choice to a strategic plan	

move into the development of strategies and strategic planning (decide and act). We applied the techniques in the fourth column to the issue of population aging in Israel in the chapters indicated.

Scan and List Initial Questions

We gathered directors-general from several ministries to use *backcasting*, presenting the title of a mock future favorable story about Israel's socioeconomic development and asking them to describe its contents. The dominant theme of the replies was the need for social and economic inclusion beyond the norm in Israel 2011.

Several participants explicitly mentioned the issue of population aging. According to Central Bureau of Statistics (CBS) projections, those ages 65 and older could go from 10 percent of Israel's 2009 population to 15 percent in 2029. In 20 years, that number could be as much as 900,000 higher than today.

We framed several initial questions at this stage:

- What changes could occur in Israel's age structure in the medium and long terms?
- What are the potential social and economic consequences?
- How have other countries thought about this, and what have they done?
- What factors will most affect outcomes of interest?
- How do these trends interact with and affect one another?
- How can goals be framed from the perspective of the elderly and the nation?
- What are some alternative strategic approaches for achieving these goals?
- How should one decide among these alternatives?
- What issues warrant the greatest attention?

Gather Information

The National Council on Geriatrics and Aging holds that, to meet future demand and maintain the current patient-to-bed ratio, the number of hospital beds would have to increase by 44 percent by 2020 and by 84 percent by 2030. In geriatric specialties, retirements are expected to exceed new entrants in the next decade.

In Israel, elderly usually remain at home, so a growing proportion of adults' time could be devoted to caring for elderly relatives. Despite this, many elderly in Israel report feelings of solitude and loneliness above the norm in other Organisation for Economic Co-operation and Development (OECD) countries. Loneliness and social isolation are increasingly seen as leading to poor health outcomes for the aged. Employment and social policies and practices that discourage work at an older age deny older workers choice over when and how they retire.

In an era of rapid population aging, they could result in underutilizing valuable labor resources.

RAND was asked to conduct a major case-study analysis of countries that have preceded Israel into demographic transition. We selected Finland, France, Italy, Japan, Korea, Sweden, and the United Kingdom. Aging-related policies and planning are ongoing in all the examined countries. Considerable experimentation occurs even within a single country with each aspect of policy showing evidence of adaptation. The inference is that population aging is best addressed with higher-level strategies and plans to which policymakers can refer in their efforts at continuous improvement.

Life expectancy of the elderly has progressively increased in recent decades. Healthy aging—i.e., maintaining the elderly in good health and keeping them independent for a longer period—is generally considered to affect the costs of health and long-term care directly, as well as independently increasing the welfare of the elderly. Policies for healthy aging could also play a role in mitigating future aging-related pressures on public finance.

Policies aimed at improving healthy aging can be grouped under four broad headings:

- improved integration in the economy and into society
- better lifestyles: focus on physical activity, nutrition, and substance use and misuse
- adapting health systems to the needs of the elderly, including in these key areas:
  - more-regular follow-up of chronically ill patients and better coordination of care
  - enhanced preventive health services
  - greater attention to mental health
  - encouraging better self-care through health literacy and access to technology.
- attacking underlying social and environmental factors affecting healthy aging.

Understand the System

The balanced scorecard in Figure S.3 can be used to develop a frame-

Table S.2
Dimensions, Metrics, and Socioeconomic Indicators Most Affected by Population Aging

Element of Vision	Ideal Metric	Candidate Measure for Indicator
Socioeconomic well-being of individuals and families	Availability of material means to achieve life goals	Household net adjusted disposable income per person
	Wealth to sustain shocks and meet goals over time	Household financial net worth per person
	Access to adequate housing	Number of rooms per person
	Housing affordability	Housing costs as share of adjusted disposable income
	Status with respect to disease	Life expectancy at birth
		Self-reported health status
	Status with respect to wellness	Self-reported limitations on activity
		Incidence of dementia and cognitive deterioration
		Rates of overweight and obesity
	Time balance between paid work, time with family, commuting, leisure, and personal care	Employees working very long hours
		Time devoted to leisure and personal care
	Capacity for and availability of lifelong learning	Share of those ages 25 and older who have engaged in further education in life skills, work skills, and general culture
	Social network support	Availability of help through social contacts and families

Table S.2—Continued

Element of Vision	Ideal Metric	Candidate Measure for Indicator
Socioeconomic well-being of individuals and families, continued	Frequency of social contact	Meet with social contacts and family at least once per week
	Water and air quality	PM_{10} concentration
	Fear of crime	Self-reported victimization
	People's overall views of their own lives	Life satisfaction
	People's present sense of satisfaction	Emotional balance
Sustainable, growing, and innovative economy	Domestic	Labor productivity
	Fiscal and monetary balance	Share of primary civilian expenditures in total government spending
Democracy, effective government, and equal access	People's perceptions of the adequacy of services	Waiting times for services (e.g., transfer from emergency room to hospital bed)

NOTE: PM_{10} = particulate matter up to 10 microns in size.

work of indicators for Israel's socioeconomic outcomes. This can then be used to identify those that might be most relevant for specific issues, such as population aging. Table S.2 provides an example of such a derivation. This framework of vision elements, goals, and metrics (that is, ideal concepts for which practical measures will need to be devised) is both a tool for analysis and a preliminary framework for active strategic and policy planning. The trends and factors affecting population aging clearly link back to the matrix in Table S.2, confirming its relevance as a priority focus from the perspective of long-term socioeconomic strategies for the nation.

Perform Dynamic Analysis

Under present trends, in 15 years, Israel's health care spending would claim a 10- to 12-percent greater share of gross domestic product (GDP) than in 2009 solely because of the shift in age distributions.

Total spending on health and long-term care could be 36 percent greater than today. The share of the population generating economic growth will also change. In 2009, every 100 working-age Israelis were matched by only 18 who were past age 65. This favorable historical circumstance is about to change: In 2024, 100 working-age Israelis could be carrying an average of 24 to 26 aged dependents in their roles as family providers or taxpayers. Not only will these changes occur over a relatively short time; they have already begun. These circumstances *will* present Israel with an inevitable "surprise."

Build Scenarios

By its nature, strategy requires thinking in time. Scenarios are narratives that follow a set of assumptions to a logical future state. Tools exist to enhance a group's ability to think systematically about the future.

RAND and Israeli government colleagues used several scenario techniques and developed a list of "what could go wrong" scenarios in the absence of policy action as Israel's population ages. This allows decisionmakers to focus attention on particulars while they gain a better understanding of how factors and outcomes, causes and effects, might be related. When we enumerated the main drivers behind the items on the list, we could see how few factors appear as drivers of several possible undesirable outcomes, as well as an underlying structure. Table S.3 shows how we can place these key factors into general categories. This also highlights areas that public policy or regulatory changes could address and those governmental actions only indirectly affect (e.g., demographics).

Identify Key Questions and Strategic Directions

For the example of population aging, at least four broad lines of action emerged from our analysis:

- **budgetary approaches** for addressing potential funding gaps. The focus is on testing and addressing shortcoming in public safety nets, encouraging the accumulation of private savings to prepare for retirement, and adjustments to taxes, benefits, and incentives to affect behavior and demand for services.

Table S.3
**Main Factors Driving Possible Negative Consequences, by
Subject Area**

Main Factor Cluster	Main Factor Driving Negative Outcome
Demography	Declining share of Israelis who are of working age
	Extended retirement years
	Insufficient labor participation
	Life-cycle patterns reducing consumption
	Specific pattern of aging in Israel
Society	Social patterns that exclude elders
	Employer and workplace attitudes
	Reliance on families for elder support
	Too few models of work–life balance
Infrastructure	Insufficient facilities for elder care
	Insufficient support for physical fitness
	Insufficient elder-enrichment facilities
Budget	Increased costs of elder care
	Competing public social welfare needs
Regulation	Regulations on health care
	Regulations on employment
	Regulations on pension and retirement
	Change in delivery to meet demand
	Regulations on education
Planning and training	Insufficient focus on retraining
	Insufficient private retirement savings
	Lack of professional financial assistance
	Insufficient training in self-sufficiency
	Lack of opt-out–only saving programs
	Insufficient incentives for geriatric-worker training

- **improving efficiencies** in cost and timing of care delivery. The government would seek improvements in data gathering and analysis; reorganization and institutional change to achieve more integrative, collaborative, and multidisciplinary care; and greater use of long-distance and in-home care.
- **broadening the base**. The focus is on improving conditions for elders to continue working before and after formal retirement, as well as immigration and family policies. Israel could broaden the base by upgrading the skills of communities with marginal participation in the general economy. This could also include changing work–life balance.
- **improving outcomes** for health and aging. This approach looks beyond the efficiency of care delivery toward enhancing its efficacy. Measures could include promoting evidence-based standards and best practice of care; seeking to enhance social interaction, inclusive participation, and support of elders; and influencing behavioral choices regarding, e.g., physical activities and eating habits.

For population aging, as with other socioeconomic issues, some normative questions could be illuminated but not resolved through analysis. Rather, political choices must be made either by default or by more-conscious balancing. For the socioeconomic consequences of population aging, some of these normative questions are as follows:

- What level of aging-related public costs is acceptable?
- What standards of care are acceptable?
- What is the appropriate mix of public and private effort and funding?
- What elements of quality of life should receive priority?
- How much economic participation should be expected from people in later life?
- What constitutes "successful" aging?

Next Steps

We have used as an example the application of a strategic perspective to the socioeconomic dimensions of aging. The next stages move beyond the sphere of analysis and into the realm of policy because they involve normative decisions and priority choices that only policy officials can make. Israel now has the institutions in place for strategic assessment. Chapter Ten of this report provides an example of how this process could be conducted within the new institutions put in place based in part on the recommendations suggested by this project (Shatz et al., 2015).

What remains is to turn this latent capacity into support for policies to achieve long-term socioeconomic strategic goals. This is a major undertaking complicated by inherent uncertainties and conflicting goals. It can be approached incrementally. What might serve the nation well is for the government to think strategically about the framework of goals and then apply its new apparatus and institutions for strategic analysis to specific issue areas. The strategic perspective will allow seeing these issues in a longer time perspective, and the framework of goals will provide a means for observing more concretely how these domains interact. The result of doing so would be greater policy choice while focusing debate on identifying a reasonable range of options.

Acknowledgments

The members of the RAND Corporation project team thank the many people in Israel who spoke with us about Israel's current socioeconomic strategy development, government decisionmaking, and the problems of Israel's aging population.

We also thank C. Richard Neu of RAND and Jack Habib, director of the Myers-JDC-Brookdale Institute, who provided insightful, detailed, and challenging formal reviews. Their comments greatly improved the manuscript while in no sense leaving them liable for any errors or misstatements that might still lurk within its pages. We take full responsibility for those.

Abbreviations

ABP	Assumption Based Planning
CBS	Central Bureau of Statistics
CO_2	carbon dioxide
DG	director-general
GDP	gross domestic product
MOF	Ministry of Finance
NEC	National Economic Council
NII	National Insurance Institute
NIS	Israeli new shekel
OECD	Organisation for Economic Co-operation and Development
OODA	observe, orient, decide, and act
PM	Prime Minister
PM_{10}	particulate matter up to 10 microns in size
PMO	Prime Minister's Office
RDM	robust decisionmaking
UNODC	United Nations Office on Drugs and Crime

WHO   World Health Organization

CHAPTER ONE

Introduction

Plans are worthless, but planning is everything.
—President Dwight D. Eisenhower
National Defense Executive Reserve Conference speech
November 14, 1957
(Eisenhower, 1958, p. 818)

The rapidity of Israel's economic growth in the past three decades has been remarkable. The economy faced years of stagnation in the wake of the war of October 1973; by the early 1980s, Israel suffered from hyperinflation that further stymied its ability to grow. Since then, Israel has been able to construct a vibrant economy with large gains in gross domestic product (GDP) per capita and has become a source of technological innovation that has positioned it as a major node in the global knowledge economy. These strengths were well manifested during the recent global downturn, in which Israel's economy outperformed that of almost all other developed countries. Israel became a member of the Organisation for Economic Co-operation and Development (OECD) in 2010.

Israel's society was a complicated mixture from the start. It has recently grown even more complex owing to the wave of immigration after the fall of the communist governments in Eastern Europe and the Soviet Union, absorption of the Ethiopian Jewish community, and Israel's own population dynamics, which have led, for example, to the formerly small ultraorthodox (*haredi*) Jewish community becoming an increasingly large share of the total population.

These powerful forces have raised numerous socioeconomic issues in Israel, including differences among the population in the willingness and ability to participate in (and benefit from) the growth in the economy, rising costs of living, and questions on the part of the public about the government's ability to address those challenges. Facing major existential challenges throughout Israel's history, the government has not routinely developed strategic responses to problems that demand longer-term, coordinated policy action.

In 2010, the government appointed a team of high-level officials charged with developing a long-term socioeconomic strategy for the state of Israel. The team, consisting of the director of the National Economic Council (NEC), the director-general of the Ministry of Finance (MOF), the director of the Office of the Budget, the deputy governor of the Bank of Israel, and the director of the Planning Division within the Prime Minister's Office (PMO), in turn, sought external support for its efforts.

RAND was asked to provide a variety of supporting inputs on such matters as the nature of strategic planning in government; issues of measurement and assessment; methods and techniques for strategic assessment and strategic plan development; appropriate institutions, processes, and data sources for enhancing capacities for strategic assessment; and comparisons with the practice in other countries. A prior publication (Shatz et al., 2015) discusses the matter of institutions, processes, and international case studies.

In this report, we highlight selected RAND inputs to the government task force. Our purpose is to provide a summary document to serve as a reference for policymakers that captures the essential mechanics of socioeconomic strategy and provides an illustrative application. We first discuss the role for strategic assessment as part of the policy planning process in Israel and then address how to define socioeconomic goals to help achieve the underlying vision. We conclude by illustrating the general points with specific examples regarding the aging of Israel's population, an issue that would benefit from a strategic perspective—not only by bridging the long- and short-term perspectives but also in easing coordination across ministerial boundaries.

Chapter Two discusses the value of providing a strategic perspective on the socioeconomic issues that Israel faces. As we discuss there, by *strategic perspective*, we mean an *orientation* designed to identify emerging socioeconomic problems and opportunities, analyze potential direct and indirect effects, and assess policies or actions to address them, as well as a *tool* for ensuring the coherence and feasibility of policies. The chapter provides definitions but also lays out what tangible benefits can come from enhancing this capacity in Israel's government.

An important component of a strategic perspective is to be explicit about visions for successful future outcomes. Chapter Three provides a general discussion of the role of visions, goals, and measures in defining what outcome a strategy is intended to produce and helping to frame the strategy itself. Chapter Four follows with an illustration applying these principles to the specific circumstances of Israel.

Chapter Five presents a schematic process for conducting an analysis intended to provide a strategic perspective on one or more socioeconomic issues. Chapters Six through Nine then present an illustration of how such a process could be applied to a specific example selected by the government panel—the issues presented by the aging of Israel's population. Chapter Six provides a frame of reference and a basic grounding in the issues through initial information-gathering. Chapter Seven synopsizes a larger report RAND provided on how developed countries with prior experience approached the challenges raised by demographic shifts, as well as a summary of a literature review and results from interviews in Israel. The chapter concludes by mapping the issues related to population aging onto the framework developed in Chapter Four.

Chapter Eight uses various quantitative techniques to conduct a dynamic analysis of trends and project possible futures, while Chapter Nine conducts a similar inquiry using scenario-building approaches. The final chapter provides a synthesis of the prior findings, points out the main strategic directions for framing policy, and discusses the steps required to carry this analysis into the process of developing a formal strategic plan and implementing it.

CHAPTER TWO

The Strategic Perspective and Socioeconomic Policy

Calls for strategy in policy abound. Developing countries are called upon to devise strategies for sustainable growth (OECD, 2001b). Security scholars and practitioners in the United States debate whether their country has lost the capacity to develop national strategy (Dubik, 2011).

Israel is not alone in its desire to enhance the strategic element in its socioeconomic policymaking. In this chapter, we discuss the concepts behind strategy and strategic perspective relevant to the development of socioeconomic strategy in Israel.

Our point throughout is that Israel will gain the greatest value if it can embed a systemic *strategic perspective* in its socioeconomic policymaking. We use this term quite consciously to distinguish the application of strategic thinking from formal strategic planning. These two concepts are intimately related but nonetheless distinguishable. Strategic planning places the emphasis on producing a particular product: a strategic plan. In contrast, a strategic perspective is an orientation, both supported by and driving a coherent set of activities and processes, designed to identify emerging socioeconomic problems and opportunities, analyze potential direct and indirect effects, and assess policies or actions to address them. A strategic perspective is also a tool for ensuring the coherence and feasibility of policies. Policies that address a specific problem or opportunity should be coherent, in the sense that they should work together and complement each other. Just as important is the requirement that they be feasible: They should take account of potential obstacles and constraints.

A strategic perspective is the precursor of and the foundation for strategic planning. The strategic perspective brings an analytical element explicitly and directly into the realm of policy decisionmaking.

In this chapter, we emphasize the following points:

- Applying a strategic perspective to socioeconomic problems is a powerful and practical tool for identifying and solving those problems over time. Israel would benefit by enhancing its capacity for doing so and by developing institutions and practices that embed such a perspective into the decisionmaking processes of the government.
- A strategic perspective helps to bridge not only the gap between the short-term focus of governments and the stewardship over longer-term outcomes that have been entrusted to them by the public but also the gaps between the ministerial portfolios that might be involved in addressing the complex problems that challenge governmental effectiveness. A strategic perspective must operate not just across time but, as importantly, across government function.
- A strategic perspective that is not systemically connected to policy implementation is less effective than it otherwise could be. Therefore, policymakers taking a strategic perspective must have a sense of the constraints and opportunities inherent in implementation, and civil servants should have input into any strategic perspective on policy and should be able to develop and exercise their own strategic perspectives.

We focus more on a strategic perspective than on the development of specific strategies. Specific strategies can be useful for a variety of reasons, including helping the public understand the government's intentions. However, specific strategies must necessarily change as implementation starts because no strategist can foresee all obstacles. In addition, specific strategies are brittle: Examination of international experience shows that policymakers can easily drop or ignore them. Indeed, elected leaders or the public might see little need to devise strategy in the government. Especially if the public responds to short-term benefits or costs, politicians might be predisposed against strategy.

A strategic perspective used to assess existing and prospective problems is harder to dismiss. If the perspective becomes widely held, it might impel policymakers and government officials to consider how their proposed actions fit together to solve a problem. For Israel, creating a socioeconomic strategy will be more than just publishing a document; "it is also about the patterns of thinking that best match resources and capabilities to achieving desired policy ends" (Drezner, 2009, p. 14).[1]

Strategic Perspective and Its Link to Strategy

Strategy is often considered to be the peak of the policy pyramid in terms of importance of its practitioners (usually the chief executive and his or her staff) and its purpose (to set the overall policy direction of government or a governmental department). Given its supposed importance, strategy and strategic thinking are often ill-defined. We require a useful, bounded definition of *strategy* to serve as a foundation for what we intend when we talk about a strategic perspective. One definition of strategy is "the calculated relationship of means to large ends" (Gaddis, 2009). An alternative is that a strategy is something that contains a diagnosis of a challenge, a guiding policy for meeting that challenge, and a set of coordinated actions to accomplish the guiding policy (Rumelt, 2011).[2]

To relate strategy to strategic perspective, we begin by presenting a basic model of decisionmaking. It was originally developed to explain why pilots of one nation can dominate the trained pilots of

---

[1] The quotation actually refers to planning, but Drezner uses *planning* in the sense that we use *strategy*. As we show, the terms *strategy*, *planning*, and *strategic planning* are often conflated or left undefined, leading to a lack of clarity in how the strategy and planning realms overlap and influence each other.

[2] For useful readings about strategy and strategic planning, besides these two authors, we recommend Mintzberg, 1994; Yarger, 2006; Friedberg, 2008; Drezner, 2009; and UK House of Commons Public Administration Select Committee, 2010.

another (Boyd, 1996). It sets decisions within a recursive process that is calibrated by updating knowledge about existing conditions and new developments. Figure 2.1 shows this process—observing, orienting, deciding, and then acting (at which point the loop process begins once more)—a so-called OODA loop.

Others have recognized that the OODA loop presents a valid model for decisionmaking within organizations, as well as at the level of the individual (Angerman, 2004). Because of the explicit emphasis on acts of observation and orientation, the application to strategy is very strong. An organization wishing to operate strategically must have an ability to scan the horizon (observe) and understand not only the factors that give rise to these observations but also what implications they have for the organization or the missions for which it is responsible (orient).

This simple model carries a variety of important implications. We focus on two. The first is that observation and orientation will both inform and influence decision and action. Second, the recursive property of this process suggests that it is a capability that must be continuously called upon. This suggests the need for observation and orientation to be integrated with the institutions for decision and action.

Figure 2.1
The Observe, Orient, Decide, and Act Loop

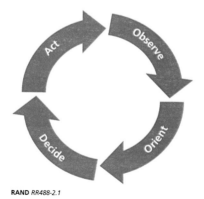

RAND *RR488-2.1*

In the context of socioeconomic strategy, taking a strategic perspective means adopting a viewpoint that considers the nature of the policy problem and the larger context in which it is placed, as well as the factors and issues that could influence any coherent set of actions that could be designed to solve it. A strategic perspective largely corresponds to the observe and orient functions of the OODA loop.

Figure 2.2 illustrates the concepts that we elaborate in this chapter and the two that follow. We have represented a series of OODA loops operating in successive years.[3] We show that this process extends from the planning period of a specific policy or strategic plan and into the period of implementation. In part, this is because this process will be engaged in at the level of a narrow, specific issue of concern to the national, general level. This allows such issues to be evaluated in a larger, strategic perspective. This recursive process takes place within the context of a larger vision for the future (the gray oval) toward which

Figure 2.2
Applying a Strategic Perspective (Observation and Orientation) Within the Framework of Long-Term Socioeconomic Decisionmaking

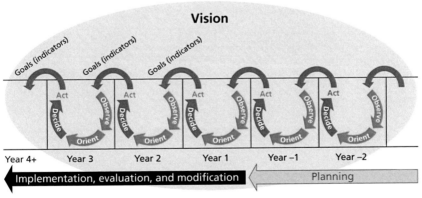

RAND RR488-2.2

---

[3] This is a convenience for the purpose of illustrating the recursive nature of the OODA loop. The actual time period for recursion might be shorter—indeed, might be close to continuous for some policy issues at specific points of their gestation and implementation. We highlight the observe and orient steps in bold for emphasis.

specific policies and strategies could be targeted.[4] In the out-years of implementation, the strategic perspective continues to apply as the actual trajectory of the policies and their outcomes are monitored by comparing goals and indicators. This creates the potential for applying modification and course correction.

Creating new institutions or charging existing institutions within the government with the tasks involved in applying the strategic perspective provides a capacity for incorporating them in a functionally meaningful way. A strategic approach to policy formation requires observing the effects of decisions and actions (observe) and then analyzing them in the context of experience, governing constraints, and new information (orient). In Chapter Five, we provide a step structure for taking a strategic perspective.

We note that the strategic perspective does not mean arriving at a single best solution. Assessment of specific strategies and selecting among them occurs in the decide stage of the OODA loop. The strategic perspective does, however, begin the process of defining the feasible options and their pros and cons in order to help elected political leaders make what they consider to be the best strategic choices. Institutionalizing a strategic perspective also implies acknowledging that strategic choices require flexibility because the nature of the policy problem, its context, and issues and other factors that could influence a solution can all change. One way to overcome the constraints of uncertainty, time, and money is to embed a strategic perspective in the way civil servants and policymakers routinely approach higher-level policy problems and even implementation tasks.

The idea of a strategic perspective is, in many ways, process-oriented. Processes, such as analysis and debate, are essential to developing, testing, and implementing strategies and strategic concepts. Likewise, capacity-strengthening can help spread the practice of taking a strategic perspective throughout the government. Processes for generating strategic perspective within and across government organizations could prove to be valuable in themselves beyond their formal

---

[4] This general theme is discussed in more detail in Chapter Three and then applied in a specific illustration for Israel in Chapter Four.

output. Indeed, these processes can help create coordination mechanisms across the government.

Israel and the Strategic Perspective

In a companion document, we assessed strategic approaches to socioeconomic challenges in Israel today (Shatz et al., 2015). We identified seven gaps in the management of Israeli socioeconomic strategy and policy. These include a lack of involvement by the political leadership, government reactivity, lack of coordination in policymaking, low levels of engagement with stakeholders, inadequate use of research and empirical findings, problems with implementing government policies, and inadequate methods of measuring and evaluating policy interventions.

Several of these findings suggest that the institutions and processes of Israel's government do not adequately provide organizationally usable observation and orientation to inform both decisions and subsequent actions. Individuals can perform these functions, but neither the functions themselves nor the resulting findings are institutionalized. This frustrates attempts to make individual insights shareable and provides no mechanism by which the insights of many offices could be developed through collaboration on particularly complex issues.

Governments around the world have responded to similar gaps in a variety of ways. We conducted case studies, which produced insights on institutions, processes, and content for the development and implementation of socioeconomic strategy. In the area of institutions, we found the following:

- Development of longer-term socioeconomic strategy benefits from the establishment of a Strategy Unit that receives the attention of the head of government.
- At the same time, institutions external to the government but closely cooperating with the Strategy Unit can provide useful input.
- Combined, not only can these institutions formulate strategic alternatives and serve as a strategic early-warning system; they

can also help foster a culture of strategic thinking throughout the government.

- Legal and procedural requirements and the attention of the head of government can create and maintain demand for strategy.
- Policy successes stemming from strategy development can also help foster a culture of strategy, thus setting in motion a cycle that could enhance demand within the government for such strategic analyses of policy alternatives.
- As an integral part of the process, strategy capabilities within individual ministries must be enhanced and can contribute not only to the functions of the Strategy Unit but also to helping demonstrate the usefulness of strategic thinking.

In the area of content and processes, we found the following:

- The Strategy Unit can have positive influence by working with a new government to create a strategy tied to the government's term. At the same time, development of longer-term strategies, strategic alternatives, and strategic early warning should be carried out as well.
- For all of these, stakeholders outside the Strategy Unit and the government should be engaged early and often.
- Strategy development and implementation can be enhanced by strengthening ministry coordination by the director-general (DG) of the PMO and by the operations of committees of DGs.
- Useful analytical tools that support strategy include budget and economic projections, research by subject-matter experts, and scenario exercises.
- Making strategy documents more widely known and available, although not possible in all cases, should nonetheless be the preferred approach not only to help build support but also to get further input for midcourse adjustments.
- Connections with the budget can aid strategy implementation, and the development of a medium-term budget framework can support strategic thinking by clarifying the means by which strategic ends will be paid for.

Informed by our findings, we and our partnering organization, Shaldor Strategy Consulting, recommended the following institutions to the government of Israel so that it can start to create a long-term socioeconomic strategy. These recommendations were presented to the cabinet in October 2012 and approved for implementation:

- Form a political-level strategy forum led by the Prime Minister (PM) to serve as the focal point for the adoption of a strategic perspective, decisionmaking regarding strategy development, and oversight of the management of the strategic process.
- Form a professional-level socioeconomic strategy forum linked to the PM to serve as the focal point for coordinated thinking and management of a new socioeconomic strategic agenda at the top civil-servant level.
- Form a socioeconomic strategy staff unit, as part of the NEC, to be the focal point of support for engaging socioeconomic challenges with a strategic perspective, strategic early warning, development of strategic alternatives, and strategy education for the civil service and to provide professional support for the Professional Forum in government agenda-formulation and management processes.
- Create a new socioeconomic strategic agenda that the government can define and promote for its term in office.
- Create a mechanism within the PMO for managing plans for socioeconomic programs selected in the government's agenda.
- Establish a deputy-DG forum; strengthen ministerial strategy, policy, and planning units; and develop ministry capabilities for taking a strategic perspective and translating strategies into policies, programs, and plans and implementing them.
- Establish a council of external stakeholders and experts to link the public to strategy processes.
- Use research, monitoring, and evaluation to assess the work of the new institutions and processes and the policies and programs that result from them.

- Create formal and informal incentives for taking a strategic perspective, participating in the new institutions and processes, and using their outputs.
- Create formal and informal incentives for taking a strategic perspective, participating in the new institutions and processes, and using their outputs.

In Chapters Five through Nine of this document, we discuss the process of strategic assessment and some practical approaches to it. In Chapter Ten, we discuss how these might map onto the institutional structure laid out in this section.

Potential Value Added by Enhanced Capacity for Strategic Assessment

In this report, we provide content for the concepts introduced in this section and suggest how the government of Israel could improve its abilities to take a strategic perspective regarding socioeconomic problems. We illustrate with specific examples stemming from the issues raised by the aging of Israel's population. Of course, these examples are not in any way to be regarded as prescriptive: Only those who have been duly placed in positions of policy authority by the citizens of the state of Israel through democratic process may provide the necessary content. But we proceed on the presumption that having practical illustrations will be of value to those officials and the staff members who support them.

Consistent with the concept of a strategic perspective, the purpose of such support to policy is

to inform and support the deliberations of top executive branch officials as they make strategic decisions. The true aim of national strategic planning is heuristic; it is an aid to the collective thinking of the highest echelons of the government, rather than a mechanism for the production of operational plans. (Friedberg, 2008, p. 48)[5]

This means, in part, better defining emerging policy problems and potential solutions; thinking about necessary actions, how they can be coordinated, and what their effects might be; and taking account of constraints and uncertainties.

Building socioeconomic strategic assessment capabilities within government will provide tangible, practical benefits to the state of Israel by saving resources and increasing the quality of government services to its citizens over time. Further, the analysis of existing gaps suggests that Israel will increasingly be confronted with serious social and economic issues for which past approaches and existing institutions might prove inadequate. Achieving this enhanced capacity requires the following:

- creating a system for an early detection of trends and opportunities
- creating governmentwide capabilities to analytically weigh potential alternative policies and their likely consequences in advance, which will increase the chances for policy success and reduce the unintended consequences of policy choices
- providing the foundation for wider discussions within the government of complex issues that are difficult to conduct on a shorter-term, crisis basis

---

[5] Friedberg was writing about national strategic planning processes in the United States, but the thoughts can apply to any institutions and processes that support the development of strategy. We would amend the quotation above to read that the purpose of the support "is not *necessarily* to produce a single, comprehensive document." That is the power of the strategic perspective: It provides considerable value in itself while constituting the necessary first steps (observe and orient, in the terminology of the OODA loop) for a well-grounded process of strategic plan development. This observing and orienting supports political leaders by describing the key trends and problems, putting them into an understandable context, analyzing the uncertainties, and building alternative future views to frame dialogue and further analysis. In doing so, they provide the tools for weighing and framing policies: identifying feasible alternatives, describing their pros and cons in terms relevant for decisionmaking, and ruling out infeasible alternatives that might otherwise have appeared attractive viewed through a less comprehensive lens.

- establishing processes that assess, on an ongoing basis, strengths, weaknesses, opportunities, and threats and feed these assessments back into strategic thinking.

The following chapters of this report are intended to serve as a resource to those seeking to construct such capabilities.

CHAPTER THREE

Visions, Goals, and Strategic Frameworks

Both a strategic perspective and any strategies that emerge from a planning process require a frame of reference. Even if the goal is only to observe and orient, there must be some way of determining what is important to focus on and how to interpret accumulated observations in a way that provides effective orientation for policy considerations and decisions. This is illustrated by placing the processes depicted in Figure 2.2 in Chapter Two within the framework of an overarching vision. In this chapter, we address this need. We begin with a discussion of visions, goals, indicators, and measures in general terms and draw from examples and practices outside of Israel.

Key Ingredient for Strategy: What Goals Are to Be Achieved?

If strategy is a tool for achieving specific ends, the identification of those ends is a crucial step for grounding strategy in the things that matter to the decisionmakers and to those in whose name they act. The first of these is a **vision**: What does a desirable future state of the world look like? The nation's leaders can help to articulate that vision or even provide their own visions, but ultimately, for the vision to have validity, the electorate of the country must widely share that vision. This need not necessarily be a formally articulated vision. We return to this issue later.

A necessary step for translating a vision into a prospectus for policy and action is to set specific **goals**. What are the parameters and levels of achievement that will be important in moving toward that envisioned future state? How attainable are these goals compared with the present state and likely future conditions? What are the public- and private-sector means for attaining them? These and other similar questions now raise the need for measurement: **Indicators** and **measures** (discussed in the next chapter) will help in understanding what progress is being made toward meeting these goals. Goals, means, and strategies might need to be recalibrated periodically in a recursive manner. As in archery, intended targets and actual points of impact might differ in practice. But having a target in mind improves one's aim.

Democratic governments face considerable challenges in defining visions, framing strategies, evaluating choices, and carrying out policies. There are several criteria for measuring success and, owing to differing visions of desirable futures, often considerable disagreement over the appropriate weight to place on each. This is all the more true when the discourse is about issues that lie in the socioeconomic sphere and unfold over long periods of time.

It is a difficult task for democratic governments to develop meaningful vision statements.[1] It might be possible to overcome the lack of explicit statements of vision by annunciating generally agreed principles that are implicit in statements that come from a country's common political discourse. Although perhaps not stated explicitly as such, these principles might be termed **vision elements** of a widely held implicit vision. Together, visions, goals, and measures and their associated indicators constitute a *strategic framework*, a concept to which we return below.

Both decisionmaking and implementation are conducted in public forums and are subject to both scrutiny and pressure from stakeholders, interested parties, and the public. In Israel, it is the exception for governments to survive their full terms. Further, governments and ministerial incumbencies everywhere tend to be of shorter duration

---

[1]   Political constitutions and similar primary documents can provide some guidance. Constitutions are written precisely for the purpose of allowing diverse groups to come together in a political union by stating the general rules that will allow these constituencies to work out solutions to the problems and needs that can arise in the unforeseeable future. They also sometimes provide overarching, guiding principles of governmental purpose.

than the time course of most strategic plans. And in the absence of crisis, there is little urgency to deal with slowly unfolding issues. Mobilization for other than short-term issues is a problem everywhere.

These complicating factors make it all the more difficult to provide some statement of the vision toward which strategic thinking is directed. The issue becomes more acute in the latter stages of the OODA loop. But as Figure 2.2 showed in Chapter Two, because the realities of government process (and contact with the real world) will require continuing repetition of the OODA process to maintain course, it is still present even if the desire is only to assess an array of issues from a strategic perspective prior to any formal statement of government strategy. It is technically possible to monitor trends and perform analyses independently of such considerations. But such monitoring and assessment will be of more practical use if they are informed by a sense of the public purpose they seek to benefit; pursued efficiently given limited resources available; directed toward areas realistically attainable through appropriate policy tools and other public means, both direct and indirect; and focused on issues that engage and will affect the public, the private sector, and those in government.

Government strategic thinking will be—and ought to be—judged on the basis of the value it delivers to the public it is intended to serve.

The Movement Toward Strategic Frameworks

A relatively recent development has been the emergence of attempts to create such explicit statements of vision elements, goals, and measures. The movement toward developing such strategic frameworks has occurred in both international organizations (such as the United Nations, the European Union, and the OECD) and individual countries (such as Canada, Ireland, and Australia) (see, for example, Canadian Index of Wellbeing, 2011; Irish Central Statistics Office, 2011; and Australian Bureau of Statistics, 2012, 2013). Several different drivers have spurred these efforts. One has been a growing movement to seek broader measures of individual and societal well-being than the sole reliance on levels and growth of GDP per capita. This has led to inquiries about the broader roots of happiness and individual sense of well-being, which, in turn, have driven the desire to develop more-encompassing accounting systems that explicitly track these components and even develop indices based on them (see, for example, French Commission on the Measurement of Economic Performance and Social Progress, 2009; and Ura et al., 2012). Attempts to construct environmental sustainability indices—leading to recognition that sustainability can be broadened also to encompass individual and social outcomes and more than just a narrow focus on growth alone—have also contributed to such efforts (see, for example, Hall et al., 2009; and Swiss Federal Statistical Office, 2012). And when an organization, such as the European Union or the OECD, wishes to illuminate specific issue areas, such as the phenomena of aging populations in developed countries, there is a tendency to accompany the policy attention with better means for definition, accounting, and measurement.

Some countries might have begun to carry on the same efforts for the purpose of improving government's ability to act meaningfully in socioeconomic areas of national interest. Quite often, this is presented as an intersection between the economy, society, and the environment (see, for example, the publications in the annual Measuring Australia's Progress series, e.g., Australian Bureau of Statistics, 2009). This is spurred, in part, by an interest in providing greater accountability for and understanding of government activities. The expectation is that this will, in turn, better prepare the political process for consensus on actions toward specific goals.

One might also see these efforts as a government's attempt to meet the unprecedented challenges it confronts in maintaining both currency and effectiveness in the face of the many simultaneous changes that, at times, appear to overwhelm a government's fundamental capacity to govern effectively. In this respect, efforts to develop strategic frameworks with explicit statements of national goals could be a way for overcoming some of the known vulnerabilities inherent in the democratic process that work against a longer-term perspective.

This consideration alone suggests an important role for such approaches in Israel. Plans that ensue might be useful in themselves. But if approached properly, the **process** by which those plans are framed, discussed, presented, and implemented could provide a considerable benefit in enhancing Israel's ability to conduct analysis and frame policy for issues that both extend over the long and short terms and cut across ministerial lines.

No country today provides a clear example of a single, integrated, and unified structure that explicitly lays out socioeconomic goals and maps a set of indicators onto those goals. But, for a strategic assessment unit within a government, or as a means of facilitating discussion among different offices within the government on strategic issues in the realm of socioeconomic policy, the act of thinking through such a framework could carry a range of benefits.

To put it another way, it will become easier for Israel to calculate strengths, weaknesses, opportunities, and threats that relate to socioeconomic considerations once it can better answer the question, "What country do we wish to be in ten to 15 years?"

Toward a Socioeconomic Strategic Framework for Israel

What might such a framework look like for Israel? What follows is but one example, and other configurations and approaches are surely possible. We have chosen to use the model of the balanced scorecard.[2] This approach produces results that both illuminate the vision component and serve as the source for deriving goals and indicators as shown in Figure 2.2 in Chapter Two.

The principal virtue of the balanced scorecard is in providing a framework that forces an integrated picture explicitly incorporating several internal and external perspectives. It is a means for articulating unspoken assumptions and translating visions and statements into a set of measurable objectives that both is integrated across all aspects of the organization and provides a complete portrait of what it is trying to achieve. We have chosen to apply this approach to Israel's socioeco-

---

[2] Briefly, this format originated in business and stemmed from the realization that paying attention solely to the bottom line could easily result in strategic blindness that could lead a business into difficulty. Instead, a company should have available ways of measuring the health of several core systems that all contribute to creating value. See Kaplan and Norton, 1992, 1996.

nomic outcomes because there, too, different perspectives and widely differing policy areas come into play. We want to create one coherent framework to play a role in all phases of the OODA loop.

Main Socioeconomic Components

Israel, like the United Kingdom, has no written constitution.[3] But it does possess a declaration of independence that sets out at least elements of a vision for the state and the society it seeks to foster. There also is a (not yet complete) set of basic laws providing some additional insight into what the elements of a balanced scorecard of socioeconomic goals might look like.[4] And the Israeli public at large supports a vibrant press in which short-term policies and events are implicitly measured against different perspectives: In other words, through their actions and opinions, what do Israelis demonstrate that they care about? These pieces might fit within a framework that incorporates several perspectives into a small set of basic components that together provide the elements of a vision for Israel's socioeconomic future.

Figure 3.1 lays out a proposed set of fundamental system components—to which we refer as *goal categories*—in a variation of the typical balanced-scorecard format. None of them is necessarily accepted by all strata of Israeli society nor is given equal weight by all. But a case based on both public declarations and past policy actions can be made for each of them to represent core concerns that should be present in a balanced scorecard.

The first large goal category is focused on aspects of socioeconomic well-being from the **individual and family perspective**. This

---

[3] Both lacunae stem from historical roots. In the case of the United Kingdom, the present structures of the state stem from a period before such practices were common or even thought to be necessary. In the case of Israel, the lack stems originally from the tense circumstances of its birth followed by a reluctance or inability to bring into the open and resolve some of the fundamental conceptual and philosophical differences among major constituent communities.

[4] Since the 1960s, Israel has passed nine Basic Laws considered to be potential components of a state constitution, but these largely address the legal authority for several core government institutions. The existing structure of Basic Laws does not yet cover all constitutional issues frequently found in the written constitutions of other countries.

Figure 3.1
A Balanced Socioeconomic Scorecard for Israel

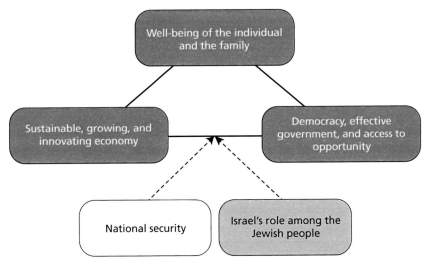

RAND *RR488-3.1*

fits with trends observed in other countries where attempts to formally raise issues of individual well-being have been made the deliberate object of policy attention.

To balance a perspective that might otherwise neglect the communal aspects of socioeconomic health, the second box, to the lower left of the first box, provides the balance of a **national perspective** on socioeconomic interests. This provides a cross-generational viewpoint as a counterpoise to what might otherwise lead to a focus on near-term satisfaction as opposed to sustainable benefit for all communities over a longer span of time.

The third category, in the box to the lower right of the first category, highlights how the architecture of a strategic framework could be tailored to incorporate particular perspectives. Conscious design choices, as much as objective principles, determine what is raised to this level of emphasis. In this instance, the chosen design explicitly emphasizes the **institutional structures** that intermediate between the individual and the wider society and economy. In Israel, there are fundamental issues that, in themselves, affect individual well-being and how an individual views his or her place in the larger society. Issues of trust, satisfaction with government services and functions, access to opportunity, and even the fundamental character of the state routinely appear in the forefront of public discourse.[5] There have been indications that such issues do have an effect on how well-being is perceived (Stoll, Michaelson, and Seaford, 2012). And because government is the main venue for public action on issues related to society and economy and is thus the locus at which a strategic framework might be of most use, it is useful for that apparatus itself to have before it a reminder of how much these institutional aspects can affect socioeconomic outcomes and the perceptions of those outcomes.

Figure 3.1 shows the first three parts of the balanced scorecard in dark colors, indicating the full force of their weight on socioeconomic outcomes. Two others have been included.[6] Because the focus is on socioeconomic goals and indicators, one could argue that the first three broad categories would suffice to contain the relevant goals. On the other hand, because one purpose of this illustration is to foster consideration and discussion, one could also argue that, for Israel, there are perhaps two other categories that require consideration.

The documents that laid the foundation for the modern state of Israel—the Balfour Declaration (Balfour, 1917), the Mandate for Palestine by the League of Nations to the United Kingdom (Secretary-General of the League of Nations, 1922), the United Nations partition resolution of 1947 (United Nations General Assembly, 1947), and Israel's declaration of establishment the following year (Jewish People's Council, 1948)—were predicated upon the vision of a democratic state for all its peoples but one that would also serve as a homeland for the Jewish people. Debate continues to this day about what each of these terms means and how they might be reconciled. But even beyond the reasons of state, the majority-Jewish inhabitants of Israel have recognized a role of stewardship for themselves that stems from their resi-

---

[5]   Participants in the backcasting exercise discussed in Chapter Six raised these issues.

[6]   One of these appears with medium shading because it will receive less emphasis than the first three in what follows, and one has no shading to signify that it will not be considered further in this report.

dence in the land that all Jewish communities see as the **center of Jewish culture and civilization.**[7]

Public policy choices made in Israel take cognizance of this additional role. Leaving this aspect of Israel's national life out of the socioeconomic equation would run the risk of presenting a structure that, from its outset, would be insufficient to reflect the trade-offs and decisions that the government actually makes. Public and private voices have aired the suggestion that, although the first 60 years of the state's existence were characterized by massive financial support for Israel by external Jewish communities, the next decades could see a reversal of trends. These same voices decry the lack of any existing strategic perspective within government from which Israel might determine what implications this could or should have for policy.

The final category, **national security**, defines the first responsibility of any national government. In most countries, except for the eternal debate over the proper allocations between domestic and defense spending, this aspect of national life can be left out of the calculus of socioeconomic strategy. This is not to state that there is no connection between the two but rather that the connection is usually indirect in most countries. However, a case can be made that, in Israel, because of its history and the role that the military serves in everything from education to professional advancement, and because the existential challenges to the state have always needed to be placed foremost in policy considerations, the concerns and actions taken on behalf of national defense have a profound influence on economic and social outcomes. In the spirit of a balanced scorecard, this aspect is at least hinted at in the design shown in Figure 3.1. It is for others to determine what role it might play but will not be considered further in this report.

---

[7] This aspect of public policy is not unique to Israel. The governments of Germany and Poland, for example, take specific cognizance of their countries' roles as the homes, respectively, of German and Polish culture. There are deliberate public policy actions that both take as a response. More than 20 European countries have some form of right-of-return law in place to ease naturalization of immigrants who can prove a link to the majority ethnicity. This role takes on a more prominent aspect in Israel partly because of the unique course of Jewish history and partly because of the heightened awareness and sensitivity that stems from unresolved communitarian issues within Israel, as well as geopolitical issues in the region in which it resides.

Dimensions Within the Balanced Scorecard

We have identified the three main accounts for scoring socioeconomic outcomes in Israel—the well-being of individuals and families, the ability to achieve sustained national economic and social development, and the effective functioning of democratic institutions along with broadly inclusive access to opportunity—as well as a possible fourth, reflecting Israel's place at the center of Jewish civilization. We need to be more concrete before we can discuss goals and the formation of strategies.

Table 3.1 arranges the three-plus-one accounting components in the first column and introduces subsidiary dimensions for each in the second.

It is a matter of judgment to determine what is a dimension in and of itself and what might perhaps be only one of several constituent aspects necessary to understand Israel's status with respect to some higher-level dimension. The decisions are based on assumptions about causal relationships among factors, as well as some statement, even if only implicit, about priorities among outcomes—perhaps even about fundamental philosophies of government. This is why developing frameworks of this type can be fundamental to strategic assessments and the processes intended to generate them. The results are of value in providing a common perspective and vocabulary; but more valuable still are the process and discussions that give rise to them.

Although many of the dimensions in Table 3.1 are self-explanatory, some are the result of design choices. Civic engagement and governance appear under the balanced scorecard's individual socioeconomic well-being account because such engagement has been found to have an effect on subjective feelings. This is distinct from the measures under the account pertaining to government performance and democratic process, which are more tuned to understanding differences between communities and locales in this dimension. Similarly, including a quality-of-life measure under democracy, effective government, and access to opportunity might appear redundant. But as we discuss below, doing so allows differences in outcomes from both geographic and ethnic community (in addition to the individual) perspectives to be addressed. This will help clarify the meaning of the similar dimen-

Table 3.1
Potential Dimensions Included Under Individual Balanced-Scorecard Accounts

Balanced-Scorecard Account	Dimension
Enhancing socioeconomic well-being of individuals and families	Income and wealth
	Jobs
	Housing
	Health status
	Work–life balance
	Education and skills
	Civic engagement and governance
	Social and cultural connections
	Environmental quality
	Personal security
	Subjective well-being
Sustainable, growing, and innovative economy	Sustainable growth
	Innovation
	Macro stability and finance
	Regulation and competition
Democracy, effective government, and access to opportunity	Political participation
	Government performance
	Access to education
	Access to employment
	Variations in quality of life
Israel's role among the Jewish people	Partnership with Jewish communities
	Connection with Jewish individuals
	Jewish culture and education

sions and measures accounted for in the individual and family well-being accounting bloc.

In this approach, we have made a design choice to erect not a minimalist structure but rather one that is richly integrated. We discuss in upcoming chapters how a leaner approach can emerge from this fundamental structure once the focus shifts to policy and implementation.

CHAPTER FOUR

Measurement and Indicators of Socioeconomic Outcomes in Israel

This chapter continues discussion of how vision, goals, and measures for socioeconomic challenges, outcomes, and policy in Israel could be integrated into a coherent strategic framework. We focus on issues of measurement and indicators of progress. As before, this discussion represents only one of many possible formulations for such a framework. But it moves us forward in considering how a strategic framework supports development of explicit statements regarding goals, as well as indicators of the state of goal fulfillment, both elements appearing in Figure 2.2 in Chapter Two showing the role of the strategic perspective in the strategy cycle. As before, we begin with a general discussion and then apply it to the specifics of socioeconomic strategy in Israel.

Metrics, Measures, and Indicators

These three terms are related but also have distinctions that are useful to keep in mind because of different roles measurement can play as part of building a strategic perspective. We use the term *metric* to refer to an ideal representation of what we seek to assay and monitor for each of the dimensions identified in Table 3.1 in Chapter Three. Each such dimension might have several metrics attached as part of the larger socioeconomic concept identified with that dimension. A *measure*, on the other hand, corresponds to the means and information we have available for actually conducting the measurement. For some metrics, such as "degree of income inequality," we might be able to construct fairly precise measures and so ensure an almost exact equivalence between the two concepts. For others, such as "level of public trust," the actual measures might be quite indirect, perhaps based on surveys or particular observed phenomena from which we draw inferences about the state of public trust. We might wish to use several indirect measures to shed light on the concept—the metric—we would want to assess.[1]

Although metrics and measures are analytical concepts, an *indicator* is closely tied to policy and priorities. Among the many measures we might identify as part of a general strategic framework for considering socioeconomic strategies and policies, we might select a subset to be indicators for assessing progress. The selection we make is consistent with the goals that the policy leadership chose for emphasis and based on their priorities. In other words, indicators are those measures to which we want to pay closest attention and on which we might base the course of policy action to be followed.

Indicators: Tools for Strategic Assessment and Strategy Development

Inasmuch as goals define what future conditions would be consistent with an underlying vision, so measures and indicators help to define progress toward those goals. As with goals and visions, measures and indicators go beyond the solely instrumental. They are more than just technocratic tools for assessing how well a particular set of government policies and actions is performing relative to a chosen strategy. They contribute an important set of capabilities to the tool kit for strategic assessment and public policy planning.

Measurement can be made integral to the processes of strategic assessment, strategic planning, policy implementation, and outcome assessment through a system of indicators. It is worthwhile to outline the ways in which such a system can influence the forms and success of public policy. Indicators are useful for **measurement**. But they

---

[1]   The government of Israel has recently instituted a procedure for developing and measuring indicators of each ministry's annual performance. A good discussion on developing metrics and measures in the context of annual ministerial work plans can be found in PMO, 2010.

also play a valuable role in conveying and supporting discussions about **ideas**, as well as in **mobilizing** political constituencies.

Measuring Progress

The most obvious role for indicators is to monitor strategies and policies as they unfold. Indicators provide mechanisms for updating information on conditions and results, which then allows for informed course correction. The process is bidirectional: Both strategies and policies will certainly require adaptation and tuning in light of actual events. Indicators can also play a large role in the type of strategic guidance documents to which Shatz et al. (2015) refers. These provide direction for the policies that will carry out a selected strategy.

Forming and Sharing Ideas

People hold different notions about how the world works. This is true even if these notions are implicit and not formalized at the individual level or within a government office. Yet, having such notions remain implicit, hidden, and unspoken is the cause of much mischief. It is important to realize the existence of these differences in fundamental ideas and their bases. This is a prerequisite for coming to a common understanding—or at least an understanding that incorporates different perspectives.

Developing a system of indicators provides a means and an occasion for staff and leaders to share ideas and concepts and connect them to visions and goals. Even more so, it provides a foundation for different government offices, perhaps responsible for different aspects of a larger, more complex issue, to interact effectively. When constructing a system of indicators, the process of indicator development can matter as much as, or perhaps even more than, the actual formal product of the process. It does so in several ways. The process of framing an explicit set of indicators provides a

- format to identify problems and provide context for their discussion
- starting point for debate or way to tie different perspectives or issues together

- step toward mutual learning, yielding a common vocabulary and sharing of values to sustain collaborative processes
- platform on which to build trust
- means for focusing on and causing explicit attention to be paid to specific trends and prospects
- foundation on which to develop strategies, as well as a basis for choosing among them
- connection between the processes of strategic assessment, strategic planning, implementation, and outcome measurement
- basis on which to construct prospective formal modeling to assess future policy alternatives.

Given the issues that every government faces in making a strategic perspective integral to its processes, this category of benefit carries the largest potential for affecting policy. Indicators are most influential when used for measurement, their first role, when there is considerable consensus on policy objectives and agreement on what types of policy are appropriate. In instances in which this level of wide agreement has not been achieved, it is in this secondary, conceptual role that indicators can have the greatest influence on strategic and policy thinking.[2]

Mobilizing for Action

Strategic thinking can be considerably more than an exercise in consciousness raising. It can provide a mobilizing force for action and can raise the chance for actions to have their intended outcomes. How such action might actually come to take place—and the way in which the larger public views both actions and outcomes—suggests a third role for indicators. They can be spurs for motivating actions and mobilizing constituencies to promote such actions.

Here again, a system of indicators can play several direct and indirect roles. As tools for building constituencies and coalitions, they can provide an entry point for wider public discussion and engagement. To this end, there can be value in inviting stakeholder involvement in the

---

[2]   Bell, Eason, and Frederiksen, 2011, provides an overview of several studies conducted to determine when and how indicators influence public policy.

development of indicators. This could enhance both their relevance and wide acceptance. Clearly, they can also be instruments for political persuasion. An indicator might prove a double-edged sword; outside of its context in a larger framework of indicators, it can take on an importance and rhetorical power that can lead to its capture and use as a political weapon.

In this light, it is clear that indicators should be well considered both individually and within a larger structure, as well as in consideration of the policy environment both inside and outside government. The institutions that frame and use the indicators must be part of the equation. As Bell, Eason, and Frederiksen, 2011, cautions, the existence of indicators does not necessarily equate to policy influence. Such influence might be limited by "low quality, low acceptance among users, limited resources, poor institutional frameworks, and disconnects between those who define indicators and those who can influence policy" (p. 13).

Constructing a System of Socioeconomic Measures

Metrics for Socioeconomic Outcomes in Israel

In what follows, we emphasize the second, conceptual role for framing a set of indicators. As such, the discussion does not include the level of detail necessary to craft a detailed measurement system. That task would be best performed after fundamental decisions about strategic direction and policy are made at the senior government level. But the resulting framework still provides entry points for basing a discussion of strategy on specific social and economic goals.

Tables 4.1 to 4.4 each elaborate on one of the four accounts in the balanced scorecard we seek to illuminate. In the left column, each table shows the dimensions that appeared in Table 3.1 in Chapter Three, while the right column presents a set of candidate metrics on which indicator measures could be based.[3]

---

[3] We selected the items that appear as candidate metrics through a process that included discussions within Israel, examination of Israeli government policy documents, docu-

Table 4.1
Dimensions and Metrics: Enhancing the Socioeconomic Well-Being of Individuals and Families

Dimension	Ideal Metric
Income and wealth	Availability of material means
	Degree of income inequality
	Wealth to sustain shocks and meet goals over time
Jobs	Availability of jobs commensurate with skills
	Quality of available jobs
Housing	Access to adequate housing
	Housing satisfaction
	Housing affordability
Health status	Status with respect to disease
	Status with respect to functional health ("wellness")

ments emerging from Israeli nongovernmental organizations and academic institutions, as well as the media, and by examining sets of metrics that had been produced elsewhere. The OECD and other international organizations of which Israel is a member have a large assortment of such documents. Examples include the OECD "How's Life?" biennial series of well-being measures (OECD, undated [a]; the OECD Labor Force Statistics database (OECD, undated [c]); Worldwide Governance Indicators (World Bank, undated); World Health Organization (WHO) health and mortality databases (WHO, undated); and United Nations Survey of Crime Trends and Operations of Criminal Justice Systems (UN-CTS) (United Nations Office on Drugs and Crime [UNODC], undated). Examples from private or nongovernmental-organization sources include the Gallup World Poll (Gallup, undated); the International Civic and Citizenship Education Study (International Civic and Citizenship Education Study, undated); European Quality of Life Surveys (European Foundation for the Improvement of Living and Working Conditions, 2012); and International Institute for Democracy and Electoral Assistance surveys (International Institute for Democracy and Electoral Assistance, undated). Other governments have both comprehensive and targeted frameworks for tracking socioeconomic outcomes. Beyond those previously cited, some examples of more-targeted frameworks would include the Early Development Instrument in Canada (Offord Centre for Child Studies, undated); the EU Statistics on Income and Living Conditions survey (European Union, undated); and the Australian Early Development Index (now the Australian Early Development Census) (Australian Early Development Census, undated).

Table 4.1—Continued

Dimension	Ideal Metric
Work–life balance	Time balance between paid work, time with family, commuting, leisure, and personal care
Education and skills	Educational attainment
	Students' cognitive skills
	Capacity for and availability of lifelong learning
	Civic learning
	Early childhood development
Civic engagement and governance	Level of citizen engagement in democratic process
	Level of trust, transparency, and effectiveness of public policy
Social and cultural connections	Social network support
	Frequency of social contact
	Social capital
Environmental quality	Water and air quality
	Environmental health impacts
	Parks and natural recreation resources
Personal security	Experience of crime
	Fear of crime
Subjective well-being	People's overall views of their own lives
	People's present sense of satisfaction

This set of metrics provides a structure for understanding specific socioeconomic outcomes and how they map onto the dimensions of the strategic framework constructed in Chapter Three. This is useful not only for identifying strategic points for analysis or intervention but also for understanding what government offices and what public and private institutions both affect and are affected by these outcomes.

Table 4.2
Dimensions and Metrics: Sustainable, Growing, and Innovative Economy

Dimension	Ideal Metric
Sustainable growth	Domestic economic growth
	Trade
	Environmental sustainability
Innovation	Innovative goods and products
	Innovative capacity growth
Macro stability and finance	Fiscal and monetary balances
	International debt and payments
Regulation and competition	Regulatory burden on product markets
	Ability to enact and enforce rules

Table 4.3
Dimensions and Metrics: Democracy, Effective Government, and Access to Opportunity

Dimensions	Ideal Metrics
Political participation	Engagement in political process, by region and community
Government performance	People's perceptions of the adequacy of services
	Government efficiency
	People's perceptions of government corruption
Access to education	Effective access to information
	Effective access to high-quality education, by region and community
Access to employment	Effective access to jobs commensurate with skills, by gender
	Effective access to jobs commensurate with skills, by region and community
Variations in quality of life	Regional and community variation in indicators for individual and family well-being

Table 4.4
Dimensions and Metrics: Israel's Role Among the Jewish People

Dimension	Ideal Metric
Partnership with Jewish communities	Involvement in Jewish people's projects
Connection with Jewish individuals	Engagement with Israel in external Jewish communities
	Attractiveness of Israel to Jewish individuals
Jewish culture and education	Israel leadership in sustaining and adding to Jewish cultural resources

Indicators and Public Policy

We have developed metrics as constituent elements of a strategic framework for considering socioeconomic outcomes that are relevant for Israel. The framework is a tool both for strategic assessment and for initiating policy development. As the process moves toward strategic planning and policy implementation, measures and indicators based on these metrics become more important in their instrumental role—as sources of evidence for comprehending the status quo, presenting candidates for selection as goals and for monitoring progress toward those goals.

Developing actual measures to use as indicators of socioeconomic development is both art and science. We touch on only a few points for doing so:

- It is important to distinguish among input measures (e.g., number of teachers), output measures (e.g., secondary school graduation rates), and outcome measures (e.g., workforce skill levels and population education levels). All three might have a place in monitoring and assessing education system performance, but ultimately the goal is satisfactory socioeconomic outcomes, so outcome measures are to be preferred.
- It might be necessary to use proxy measures if the desired outcome is not directly measurable or data are unavailable. That being said, care must be taken that the selected measures provide a sufficient representation of the issues involved so that there can be reason-

able assurance that policy changes based on observation of these indicators will be moving in the desired direction.

- Where measures based on empirical data do not exist, it might be possible to engage in modeling that can be used to provide a sense of bounds, degrees of freedom, and variability in possible outcomes from policies being considered.
- Producing indicators is not in itself sufficient for achieving policy influence if the data are recognized as poor, the construction of indicators viewed as faulty, or the indicators fail to gain acceptance or are otherwise viewed as invalid by the policymakers and other intended users of the indicators.
- Comparative studies of indicators and policy within the European Union suggest that, if they are intended for use in their conceptual role to inform the public, simpler and more-aggregated indicators are most effective. Composite indices are prominent in this role. To fulfill the instrumental role of guiding policy and allowing for measurement of policy effects, more disaggregated information is more acceptable and influential (Bell, Eason, and Frederiksen, 2011).
- Indices of composite indicators might have greatest use in raising public awareness. Policymaker demand appears to be low (Bell, Eason, and Frederiksen, 2011).
- Sets of indicators should be reviewed over time for their continued relevance. The very popularity of a particular indicator could become problematic if it no longer accurately reflects the actual conditions it is intended to measure.

Indicators of Socioeconomic Outcomes

Appendix A provides an illustration of how measures within a strategic framework might form a set of potential indicators for socioeconomic outcomes. For each ideal metric, we provide an example of one or more actual indicator measures. For several, we indicate potential weaknesses, as well as reference to the use of a similar indicator elsewhere.

As can be seen, the list varies greatly in terms of measure format, direct or indirect measures, reliance on survey or time-series data, and other elements. There might also be differences in the frequency with which they might—or can—be revised. Doing so on a regular basis would provide a dashboard of indicators of socioeconomic outcomes in Israel.[4]

With the goal of being comprehensive, the length of the list is potentially daunting. In several instances, more than one measure is associated with a single ideal metric. In some cases, it could be that no single measure is practicable and so several approximate measures, perhaps weighing different aspects, are included. Even in such instances, not all need necessarily be utilized. Neither is it necessary to assign equal weight to each. In fact, a proposed set of indicative measures, such as this, should be viewed as a menu. Determining which are selected for emphasis as indicators will depend on several factors. Certainly, the relative ease with which each indicator can be constructed is one of these factors. For some, the data are already routinely collected. Others might require greater effort or might be judged as not sufficiently illuminating to warrant the labor and resources involved.

Even if each indicator could be constructed with similar ease, there still would be reason to be selective about the smaller number to be actively tracked. Which are chosen and how they are used are important policy choices and should be regarded as such. In this sense, a set of indicators can provide a tool for constructing the types of documents prepared by the strategic planning and policy forums discussed in detail in prior work for this project (Shatz et al., 2015). A strategic framework, such as in Appendix A, is an active tool for the following:

- encouraging comprehensive thinking about and observation of a broad range of socioeconomic issues and outcomes in Israel
- acting as a basis for defining what might be acceptable ranges of values for each indicator or ideal metric, both of which are ultimately related to dimensions and goals

---

[4] Analogous to the dashboard instruments of an automobile, *dashboard* in this context would refer to a set of measures chosen to be a small set of indicators for monitoring the results from policy initiatives and socioeconomic outcomes.

- providing structure for the horizon-scanning (observation) and early warning (orientation) that are important elements of the strategic perspective OODA loop
- serving as a means for defining and constructing forecasts or framing scenarios.

CHAPTER FIVE

Applying a Strategic Perspective

In prior chapters, we introduced the concept of a strategic perspective, as distinct from strategies or strategic plans, with emphasis on a continuous process rather than distinct products. Analogizing from the OODA loop familiar to graduates of Israel's officer corps, a strategic perspective explicitly combines processes of observation and orientation to inform government decisions and guide government actions. We advanced a possible balanced scorecard framework for directing active steps of observation and then orienting such observations with reference to Israel's socioeconomic agenda. Finally, we suggested one possible system of measures and indicators for socioeconomic outcomes in Israel.

How can these concepts help determine where opportunities or threats to socioeconomic well-being might lie? In this chapter, we suggest a protocol of steps for organizing such an assessment. In doing so, and in the balance of this report, we emphasize the early-warning and initial issue-framing roles for the strategic perspective. Whereas, in Chapters Three and Four, we discussed the higher-level elements of Figure 2.2 in Chapter Two, in this chapter, we look inside each iteration of the OODA process, focusing on what specific analyses might support the actions of observation and orientation. To provide examples of how these steps can be performed, we apply them to a multifaceted issue driven by underlying forces largely beyond the time scale of daily events and so not readily visible.

Steps in Applying a Strategic Perspective

It is convenient to present the process of developing a strategic perspective as a sequence of steps. In practice, the actual transition between steps and the sequence in which they occur can both become considerably more complicated. Especially if a Strategy Unit must look across many potential issues, some of the later steps, once performed, might shed more light and lead to revision of those that had come before. There might also be precipitating events or unusual opportunities for access to insights or information that cause a strategic perspective to be instigated at a later stage in the process. The prior stages might then be developed only in retrospect. However, precisely because daily practice can involve compromises, it is useful to lay out explicitly an idealized view of how a strategic process might logically proceed.

Table 5.1 shows the major conceptual steps involved. The first column identifies the portion of the OODA loop in which the step takes place. The second column lists the steps, and the third provides some explanation of their purpose. The first six steps (rows) are sufficient for developing a strategic perspective. The last three steps would take the results of the first six and move more directly into the development of strategies and strategic planning, the decide and act portions of the OODA loop. We focus on the observe and orient steps through Chapter Nine and then, in Chapter Ten, consider how the decide and act steps might fit within the institutional framework outlined in Chapter Two.

Table 5.1 presumes that there has already been some statement of vision and overarching goals on the part of the policy leadership within a strategic framework, such as was presented in Chapters Three and Four of this report. It then details a course of analysis within that set framework once a particular issue or socioeconomic outcome has been selected for focus. The highlighted six initial steps correspond to the OODA processes of observing and orienting. A **general scan** of status and trends will raise questions for follow-up. **Information-gathering** emphasizes the utility of reaching widely inside, as well as outside, government for expert input once some of the priority questions are designated for follow-up. Thinking through **systemic con-**

Table 5.1
Steps in Implementing a Strategic Perspective in Decisionmaking

OODA Stage	Step	Purpose
Observe and orient	1. Scan and list initial questions.	Looking forward and identifying potential issues for focus
	2. Gather information.	Making use of prior work and available information to create a base of current initial information
	3. Understand the system.	Identifying key factors and forces and how they interact within the socioeconomic sphere
	4. Perform dynamic analysis.	Exploring trends, uncertainties, and possible trajectories
	5. Build scenarios.	Constructing alternative future views designed to frame dialogue and analysis
	6. Identify key questions and strategic directions.	Synthesizing and integrating, which could motivate more-detailed analyses and preparation for action
Decide and act	7. Define strategic objectives.	Deriving and achieving consensus on a set of desirable outcomes
	8. Assess alternative strategies.	Examining strategic choices systematically and building a robust course of action
	9. Plan the implementation.	Determining requirements for moving from a strategy choice to a strategic plan

nections helps make clear the importance of these questions, and the answers received, to the goals and priorities of government and the public interest: Which of these will be most affected? What other areas might be affected indirectly? What indicators are important to watch for developments?

Dynamic analysis—the quantification of trends and possible causal effects, along with degrees of uncertainty—begins the preparations for identifying and deciding among alternative courses of action. This should be conducted along with **scenario analysis**. These can certainly be made quantitative by using models when available. But there is value in carrying out more qualitative scenario exercises to engage relevant parties within and outside of government, to expand consideration of implications and possible courses and outcomes, and as a basis for the preliminary identification and development of possible alternative courses for action. (No one method is likely to prove either indispensable or sufficient. The methodology should not be the focus at the expense of the process it is intended to support.)

The next step would then synthesize all that has gone before by placing the results in the context of the overarching vision and broad government goals based on the strategic framework, presenting several possible **strategic directions** for consideration (including assessment of potential strengths and weaknesses of each), and framing the **key questions** that those conducting the planning and analysis need to bring to policymakers' attention.

Population Aging in Israel: An Inevitable "Surprise"

The Israeli senior government task force on socioeconomic strategy identified an important socioeconomic issue to be used as a laboratory for demonstrating strategic assessment ideas and methods, as well as providing a model for other socioeconomic issues that future governments might consider. Its final decision was to focus on the issue of demographic change brought about by the aging of Israel's population.

This topic highlights the practical benefits of applying a strategic perspective to policy. Population aging could be termed an "inevitable surprise"—a quiet, gradual, but nonetheless profound change that might not be anticipated before it manifests itself as a crisis. If recognized too late, several policy options that might have been employed in advance are no longer available. A government lacking sufficient foresight would be left with a limited number of unsatisfactory alternatives.

Table 5.2 reproduces the last two columns of Table 5.1 in its first two columns. The last column of Table 5.2 shows how several of the inputs we provided to the government team's discussion of the aging issue, discussed in detail in the chapters indicated, would map onto this idealized sequence. In fact, we could easily place several of these

Table 5.2
Strategic Perspective Applied to Population Aging in Israel

Step	Purpose	Aging Analysis Example
1. Scan and list initial questions.	Looking forward and identifying potential issues for focus	Backcasting exercise (Chapter Six); demographic trends in Israel (Chapter Seven)
2. Gather information.	Making use of prior work and available information to create a base of current initial information	Subject-area overview (Chapter Seven); general research on aging (Chapter Seven); international experience (Chapter Seven)
3. Understand the system.	Identifying key factors and forces and how they interact within the socioeconomic sphere	Map onto indicators (Chapter Seven); select priority indicators (Chapter Seven)
4. Perform dynamic analysis.	Exploring trends, uncertainties, and possible trajectories	Health-cost trend analysis (Chapter Eight); dependency ratio analysis (Chapter Eight)
5. Build scenarios.	Constructing alternative future views designed to frame dialogue and analysis	Futures wheel (passive) (Chapter Nine); futures wheel (active) (Chapter Nine); "what could go wrong?" (Chapter Nine)
6. Identify key questions and strategic directions.	Synthesizing and integrating, which could motivate more-detailed analyses and preparation for action	Defining main questions (Chapter Ten); strategic alternatives (Chapter Ten)
7. Define strategic objectives.	Deriving and achieving consensus on a set of desirable outcomes	
8. Assess alternative strategies.	Examining strategic choices systematically and building a robust course of action	
9. Plan the implementation.	Determining requirements for moving from a strategy choice to a strategic plan	

inputs into multiple rows of the table. As can be seen from this third column, we were not requested to go beyond the strategic perspective steps. Actual strategy development is the responsibility of Israel's policy authorities. It would not be appropriate, even if only in example, to imply that this is something that is in the province of the strategic planners alone. This division also makes clearer where and how a strategic perspective or strategic assessment might be assisted by work performed outside the government as distinct from later, strategic planning steps for which such involvement might raise questions about potential undue influence.

The balance of this report presents a selection of the inputs that the government task force requested from us. We utilize portions of these inputs to illustrate what could flow from a strategic perspective process; the inputs cannot themselves provide a single narrative or a full exposition of this issue area. By no means do these represent a complete assessment, but they do illustrate how several of the items that the task force requested in 2011–2012 would fit within the framework we propose in this chapter.

In this light, we present several insights into population aging in Israel and the potential influence in socioeconomic outcomes. We do so in several stages, starting in Chapter Six with *scanning and listing initial questions*, as indicated in Table 5.2. In this first stage, we seek to understand the phenomena associated with population aging, identify parts of the socioeconomic system that might cause or be influenced by potential effects of this demographic shift, and frame the appropriate questions that need to be considered. We start with a method that we applied generally to socioeconomic challenges and then move to focus specifically on aging.

In Chapter Seven, we continue the *information-gathering* begun in Chapter Six with more focus on Israel, as well as international experience. We conclude this chapter by mapping aging issues onto the indicator framework we provided in Chapter Four to present a means for *understanding the system*. We next use various methods for conducting *dynamic analyses* in Chapter Eight to identify important trends, forces, uncertainties, and possible trajectories. We base these efforts on trend analyses derived from projections of time-series values. In Chapter Nine, we rely more on a few simple *scenario-building* approaches for wider examination of alternatives.

The concluding Chapter Ten moves back to a more comprehensive view as illustrated in Figure 2.2 in Chapter Two. Whereas, in Chapters Six through Nine, we focus on some specific methods for implementing the observe and orient portions of the OODA loop, in Chapter Ten, we present several ways to begin identifying and assessing strategic alternatives for Israel—that is, moving into the realms of decision and action. These are the factors that could then feed into the appropriate strategic and policy authorities to become inputs into the process of developing a strategic concept for one or more government courses of action.

CHAPTER SIX
Scan and List Initial Questions

We begin at the broadest level. In fact, the first project effort was to engage senior ministry officials throughout the government in an exercise designed to broaden perspectives and begin identifying socioeconomic issues of strategic importance to Israel. This workshop exemplifies a technique that can be useful in the early stages of strategic assessment. It also began a process of framing the aging issues within the context of a larger set of socioeconomic concerns.

Backcasting

Effective long-term planning by individuals and organizations is often trumped by pressing near-term demands for their attention and resources. Further, the sheer number of possible futures makes it difficult to determine which might be most worthy of consideration today as motivation for possible short-term action. How can such thinking be relevant for government agencies and policy bodies responsible for socioeconomic well-being? As a practical matter, it is usually difficult enough to find sufficient time and attention to deal with day-to-day concerns so as to preclude any serious consideration of the future.

Several techniques help groups stage structured consideration of potential futures as a way to bring more focus to issues of concern and possible courses of action for the shorter term.[1] In November 2011, we

---

[1] The scenario methods that Peter Schwartz and others developed and popularized are tools sometimes used for such purposes. See Schwartz, 1991; and Van der Heijden, 2005, for

engaged DGs from several ministries in a backcasting exercise. Backcasting is designed to allow a group to reverse the usual flow of thinking about futures. Rather than begin from the present and attempt to reason across multiple futures (a task made more complicated by the demonstrated poverty of human imagination to conceive of truly distinctively different, as well as analytically meaningful, futures) (Lempert, Popper, and Bankes, 2003), backcasting presents a particular future state of the world and asks participants to reason backward from that point to the present.

In this workshop, we presented senior government officials with a mock front cover from a prominent international economics newspaper dated some 15 years in the future. The cover story was "'*Altneuland*': Why Israel Now Works."[2] The cover further proclaimed, "Inside This Week: A 14-Page Survey on the 'New' Israel." We placed the DGs and other invited participants into groups of four to six and asked them to address two basic questions:

1. What is the content of the story about Israel in the year 2027?
2. What story might describe the path of how Israel arrived at that point 15 years from now? (What was necessary for these changes to happen? What did *not* happen so that the changes could occur?)

We asked each group to confer and prepare responses. Once it had done so, we presented each group with a set of paper slips. We asked each group to choose two blindly. On each slip was a wild card: some surprising (in the sense of trend-breaking) event posited to have occurred at some time between the present and the end date of the backcasting exercise. After consideration and discussion, we then asked

---

prominent examples. These are best utilized when addressing specific areas of focus rather than in a mode of discovery, as was our purpose in the exercise we describe.

[2]   *Altneuland* [Old new land] was the title of a novel by Theodor Herzl, the founder of modern Zionism, set in a future, idealized Jewish state (Herzl, 1902).

each group to report its initial answers to the first two questions, as well as how it would respond to two related further questions:

3. How do the wild-card events affect the group's original story of the path Israel took to achieve the desirable future in the story?
4. How would the wild-card events affect the likelihood that the future positive story about Israel would actually be written?

The backcasting exercise served several purposes. For the workshop organizers, it provided several types of information, both direct and indirect. Among the former are insights into what constitute widely held elements of a vision for desirable futures for Israel, how government leaders believe the socioeconomic sphere operates in Israel and is affected by domestic and international factors, and thoughts on what strategic leverage points might bring about a desirable socioeconomic future for Israel. We incorporated these findings into several of the outputs described elsewhere in this report.

Appendix B includes some selected responses to our backcasting workshop. The dominant theme was the need for social and economic inclusion beyond what was perceived as the norm in Israel 2011. This was not surprising because the workshop occurred only a few months after the large demonstrations protesting cost of living and perceived unfairness in burden-sharing occurred in Israel's major cities during that summer. Participants expressed the theme mostly in terms of closing various gaps, particularly by enhancing socioeconomic outcomes for those currently unable to realize in their own lives the prosperity experienced by Israel as a whole. It is also notable that participants made reference to indicators and measures that would be applied at the individual level. This is in contrast to the aggregate measures of well-being more typically used to measure socioeconomic outcomes in Israel heretofore.

The indirect insights were more diagnostic in nature. The workshop allowed observation of dynamics of the groups to better understand how conversations about the future take place within Israel's government. To be sure, the workshop differed from the normal course of discussion about public policy. But it did, nonetheless, suggest both the barriers and opportunities that exist in making such topics more explicit. It brought out, for example, the connection that exists in the minds of government officials in Israel between such terms as *future* and *strategic* on one hand and semantically negative words (in the sense of being in contrast with practical application), such as *theoretical* and *academic*, on the other.

For the participants, there are also several potential direct and indirect results. Of course, the main direct result can be an expanded frame of reference providing more of a future focus that can, in turn, provide more insight into shorter-term, more day-to-day planning and decisionmaking. This is the ultimate purpose of any scenario-based exercise. In particular, the two-stage structure employed in the November 2011 workshop, in which surprises are introduced into what otherwise would be a process of linear extrapolation, introduces the concept of robustness into thinking about potential strategies for achieving desirable future goals. A more extended exercise could be used to discover which potential strategies are more or less vulnerable to surprise, the assumptions on which such strategies are based, and how specific strategic courses could be engineered to enhance the chances for successful outcomes by reducing vulnerabilities to foreseeable (and even unforeseen) future challenges.[3]

The indirect benefits have the potential for being at least as useful. They shift the focus from the actual exercise outputs to the process involved for carrying out the exercise. Clearly, such process benefits would be much more likely to accrue from a more extended and in-depth effort than from a one-off exercise, such as the workshop we are describing.[4] But even a single exercise can develop channels for communication that might otherwise not exist or be underutilized. The exercise could give rise to a shared vocabulary for further discussion or even a more widely shared vision or conceptual model for both under-

---

[3] Methodologies, such as assumption-based planning (ABP), address the identification of vulnerable assumptions and enhancing robustness in strategic planning. See Dewar, 2002; and Lempert et al., 2008, for examples.

[4] Examples would include the Project Foresight process conducted in the United Kingdom in the 1980s and some more-recent efforts conducted under the auspices of the Asia–Pacific Economic Cooperation's Center for Technology Foresight.

standing the forces at work and developing appropriate responses. It might be the case that the next time the individuals or units that participated in this effort need to interact, participants might have an enhanced common basis for those conversations to occur and the conversations might be more likely to proceed in fruitful directions.

Some of the themes brought out in the workshop would justify a focus on the demographic and economic trends associated with population aging. In most societies, the elderly population has the potential for being among the most vulnerable and so, although not necessarily named explicitly, would fall into the categories of greatest concern that this process elicited. As we discuss, the basic demographic, economic, and social forces associated with population aging, if neither recognized nor addressed by examination and modification of current policies, could well lead to phenomena that would move Israel away from the vision for Israel 2027 expressed in this workshop. Therefore, addressing these issues could serve to prevent deterioration in several measures of importance and might even serve as points of leverage for moving closer to that vision for a future Israel.

In this light, it is interesting to note that, even in this broadest of contexts, in which literally everything associated with economy and society in Israel could be placed on the table, some participants explicitly mentioned the issue of population aging. This occurred twice: Once in the context of utilizing the potential addition in human capital represented by increasing life expectancy, and once in the context of the social and psychological issues associated with aging that would need to be addressed if quality-of-life indicators are applied to measure movement toward the envisioned future Israel.

Assessing Information to Establish Initial Questions

What are the basic elements of population aging in developed countries, and how might these affect future socioeconomic well-being in Israel?

It is well understood that advances in health have led to reductions of infant and childhood mortality rates since World War II. This has led to a boom in world population generally and tremendous growth in less-developed economies in particular. As more healthy children survive into adulthood, this has a profound effect on average rates of life expectancy.

Age-specific mortality rates have also changed, particularly in the age categories of those who have usually retired from the labor force in developed economies. The presumption has been that, although medical science has made tremendous strides, there would be a natural point of diminishing returns. This might be an assumption that needs to be revisited. Figure 6.1 shows the highest average female life expectancy by year recorded in the country that has the highest average in that year. The resulting linearity is quite striking. With great regularity, this highest average life expectancy has increased by three months for each year since 1840. The most interesting part of the figure for the present purpose is the horizontal lines representing prior attempts to derive the expected maximum for life expectancy. At least 11 of these presumed ceilings have now come and gone with no break in the trend.

We can infer the potential effects on public policy by noting that, in 1935, when the Social Security old-age pension was introduced in the United States, the age for retirement with full benefits, 65 years, was equal to the approximate average female life expectancy in the highest-recording country during that year. By 2010, that retirement age had remained unchanged across three-quarters of a century. Yet, it had fallen more than 20 years behind the highest recorded average life expectancy.

Two other factors affect the bearing that these demographics have on policy and the strategic choices Israel faces. The first of these are the indications that extended life spans need not necessarily mean more years of ill health. This is an additional example of a prior assumption being reconsidered on the basis of emerging evidence. A more complex equation is taking form suggesting that, with increasing age, nongenetic factors that affect longevity could come to match genetic factors in terms of importance. Among those nongenetic factors are a person's levels of physical activity, social engagement, mental stimulation, and even increasing years of labor. Some data suggest that high

Figure 6.1
Average Female Life Expectancy at Birth in the Country Recording the
Highest Level, by Year, 1840 to Present

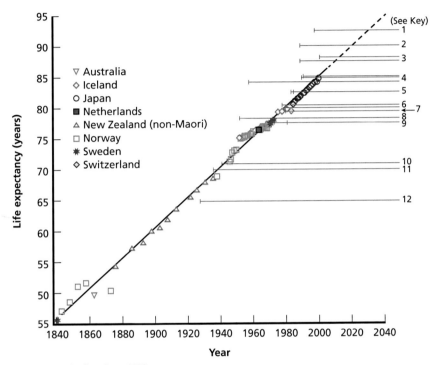

Key: 1. United Nations, 1999
      2. Demeny, 1984
      3. Olshansky, Carnes, and Désesquelles, 2001; United Nations, 2001
      4. Fries, 1980, 1989; Olshansky, Carnes, and Désesquelles, 2001; Coale, 1981;
          Coale and Guo, 1990
      5. Demeny, 1984; United Nations, 1986
      6. Bourgeois-Pichat, 1952, 1978; United Nations, 1973
      7. Siegel, 1980
      8. Bourgeois-Pichat, 1952, 1978
      9. United Nations, 1986; Frejka, 1973
     10. Dublin, 1941
     11. Dublin and Lotka, 1936
     12. Dublin, 1928

SOURCE: Derived from Oeppen and Vaupel, 2002.
NOTE: The solid slope indicates the linear regression. The dashed line indicates the
extrapolated trend. The horizontal gray lines show asserted ceilings on life
expectancy, with a short vertical line indicating the year of publication for each.
RAND *RR488-6.1*

levels of engagement in each could serve to stave off the onset of ill health and disability related to age by significant numbers of years.[5]

The net effect of this complicated pattern of interwoven trends can be seen in Figures 6.2 and 6.3. Figure 6.2 reports mortality rates and self-reported assessments of health, by age, in the United States for the recent past and a similar period three decades earlier. Not only is the decrease in age-specific mortality clear (those age 67 in the later period having the same mortality rates as those age 60 in the 1970s); there is also a trend of people reporting the same degree of health as those who would have been nine or more years younger 30 years before did.

Figure 6.2
Mortality and Self-Assessed Health in the United States, 1970s and 2000s

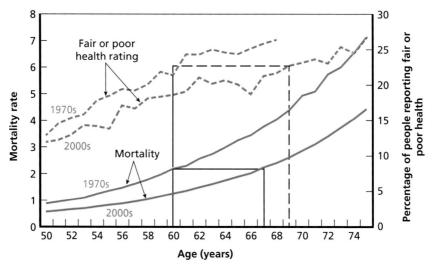

SOURCE: Wise, 2012, p. 19.
RAND RR488-6.2

[5] Rohwedder and Willis, 2010, found direct effects on cognitive ability from early retirement. Degree of social engagement also affects health and health outcomes (Luo et al., 2002; Steptoe et al., 2013).

Figure 6.3
Changes in Mortality and Self-Assessed Health for Men, Ages 60 to 64

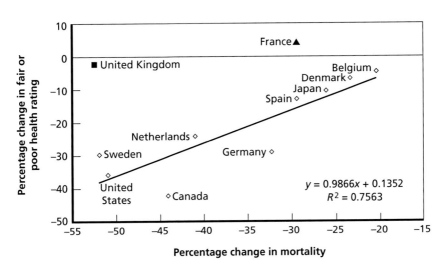

SOURCE: Wise, 2012, p. 19.

RAND RR488-6.3

This trend holds true for many developed economies, as can be seen in Figure 6.3.[6]

Characteristics and conditions we have associated with aging might be subject to modification by deliberate actions. Whereas previously, it might have been broadly accurate to think of a modern society as being made up of the young, the working-age, and the old, it might be more accurate and more useful to think in terms of *four* age classes for policy purposes: the young, the working-age, the retired, and the oldest old. Each has distinct characteristics and implications for its contribution to and need of public services.

The other factor that differs from prior experience is the near-universal decline in fertility rates among developed countries. Several

---

[6] The differences in mortality and self-assessed health are measured across different time periods for different countries. In each case, the comparison is between the most-recent data on self-assessed health with the earliest (e.g., for the Netherlands, the years are 2008 and 1983). Mortality data were selected to match the same set of dates for each country. Both France and the United Kingdom used scales for measuring health that differed from those in the other countries reported. Therefore, the data are not strictly comparable.

countries in Europe and Asia have now fallen below the replacement rate of 2.1 children per woman. This means that, in some countries, populations have begun to shrink.

The combination of these three new trends—increases in life expectancy on average, longer (and potentially not markedly unhealthier) lives, and a decline in fertility and births compared with deaths—combine in such a way that, taken together, this transformation of traditional population age profiles raises a series of questions:

- Can older people continue to be supported in their old age?
- What will be the economic, social, and political relationships between generations as the degree of overlap between them grows?
- What is government's role in preparing for change of this magnitude at both the individual and public levels?

Questions such as these have called attention to the phenomenon of population aging as a strategic issue. Although many demographic trends have eminently predictable aspects (for example, all people who will turn 65 in the year 2035 are alive today), the effects of these foreseeable trends take years to develop to the point at which they become noticeable. By that time, however, many options that could have been taken at earlier times to prepare and perhaps ease these potential effects will no longer be available. The strategic perspective requires that these trends be recognized today in order to consider courses of actions in the short term that could have profound consequences for how well a country weathers this transition.

What Are the Strategic Questions for Israel?

It now becomes possible to frame some of the main strategic questions that Israel needs to address as it determines the extent to which population aging should assume priority in policy discussions. These questions include the following:

- What changes might occur in age structure in the medium and long terms?
- What are the potential social and economic consequences of such change?
- How have other countries thought about this, and what have they done?
- How do these trends interact with and affect one another?
- What factors will most affect outcomes of interest?
- How can goals be framed from the perspective of the elderly and the nation?
- What are some alternative strategic approaches for achieving these goals?
- How should one decide among these alternatives?
- What issues warrant the greatest attention?

The following chapters do not address all the questions, both because of the cursory nature of their analyses and because several of the questions arise only in the later stages presented in Table 5.2 in Chapter Five. But we do illustrate how a framework could be established for providing the answers.

CHAPTER SEVEN

Information-Gathering and Mapping the System: Israeli and International Experience with Population Aging

The *information-gathering* stage lays a foundation for developing a strategic perspective in an area of policy attention. In this chapter, we give a brief overview of Israel's demographic trends, provide results from literature searches and interviewing in Israel, and explain how countries further along the path of demographic transition have framed policy for aging and aging-related issues. At the end of this chapter, we demonstrate how such information can then be used to map aging-related issues onto the framework for measurement developed in Chapter Four as a means for *mapping the system*.

Demographic Dynamics in Israel

Israel is confronting the same trends and forces as in other countries but with some potentially important differences. There is no question that longevity trends are already at play. Israel has the third-highest male and tenth-highest female average life expectancy in the world. Unlike most other countries in the OECD, it still enjoys overall fertility rates above replacement. But the situation is changing. Some of the factors that kept Israel a comparatively young country (a postindependence domestic "baby boom" combined with an influx of relatively young immigrants) now contribute to a large growth in the 65-and-over age cohort with the passage of time. Some groups in Israel that had not previously experienced the phenomenon of having a relatively large percentage of the population being old, particularly the ultraorthodox and the Arab communities, now face that prospect, as does the country as a whole.

The Central Bureau of Statistics (CBS) created projections of population growth out to the year 2059. Because of the changes that relatively small differences in the underlying rates would make when looking out two or more generations, the CBS created three base sets of assumptions for low, medium, and high levels of growth. The exercise is valuable because it makes clear the relationship between what is knowable and what remains uncertain. One view of these projections is presented in Figures 7.1 through 7.3, which first show the 2009 age pyramid and then projections out 25 and 50 years.

In each case, we have applied the medium-growth assumptions to the three main communities being projected: Jews and others, excluding *haredim*; the *haredi* community; and Arab citizens of Israel.[1] The blue whiskers flanking each bar show the range of possible projections within the fundamental medium-growth scenario assumptions. The uncertainties grow over time, especially for the youngest age cohorts. Two further trends are clear. One is that the pyramid becomes less pyramid-like as time passes and the 80-years-and-older cohort grows larger. The second is that, under the assumptions of this medium-growth projection, there is considerable growth in the *haredi* and Arab sectors compared with that represented by non-*haredi* Jews and others. As we discuss in more detail in Chapter Eight, this factor could prove significant if members of these two communities continue to be less economically engaged than those from the majority—non-*haredi* Jewish sector.

The bottom line from these data and assumptions is that, in the next two decades, growth in the elderly population in Israel is likely to result in an 80-percent increase in the number of people who are 65 and older while the general-population increase under the same assumptions will be only 32 percent (Chernichovsky and Regev, 2012, p. 556). We note in passing that one must not preclude that the future

---

[1] The low-growth assumptions would, obviously, lower the total population size while creating a greater proportion of older age cohorts than of young ones.

Figure 7.1
Age Structure of Israel's Population in 2009, by Community

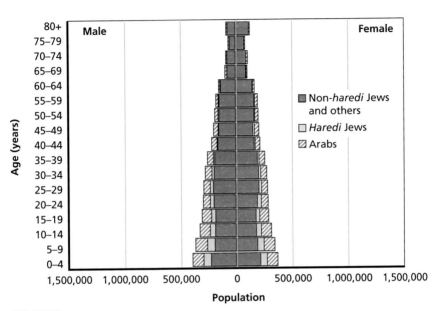

could witness unforeseen immigration waves, such as that of the 1990s, which saw an influx of Russian immigrants.

Interviews and Existing Literature on Aging Issues in Israel

Continuing with the exercise in data-gathering, we now present prominent themes from a survey of literature, supplemented with local interviewing, conducted in 2011–2012. We have arranged the information by topic.

Health Care for the Elderly
Hospital Infrastructure
Although changes in care have caused a general decline in numbers of beds throughout the developed world, the number of hospital beds in

Figure 7.2
Age Structure of Israel's Population in 2034, by Community, Medium-Growth Assumptions

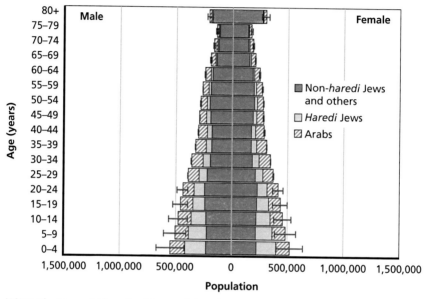

NOTE: The blue whiskers flanking each bar show the range of possible projections within the fundamental medium-growth scenario assumptions.
RAND RR488-7.2

Israel remains below those of the rest of the OECD on a per capita basis (Chernichovsky and Regev, 2012). The growth in the relative share of elderly in the population will increase the need for beds. The National Council on Geriatrics and Aging estimated that an increase of 44 percent would be required by 2020 and 84 percent by 2030 to maintain the current patient-to-bed ratio (Stessman, 2011).

Geriatric and Health Care Staffing

There are indications of shortages of health care professionals, especially in geriatric specialties, in which retirees are expected to exceed new entrants in the next decade.[2] Even though the number of geriatric

---

[2] As a result of the low number of beds per capita, the average hospitalization period in the Israeli central medical centers has shortened and is now one of the shortest in the developed

Figure 7.3
Age Structure of Israel's Population in 2059, by Community, Medium-Growth Assumptions

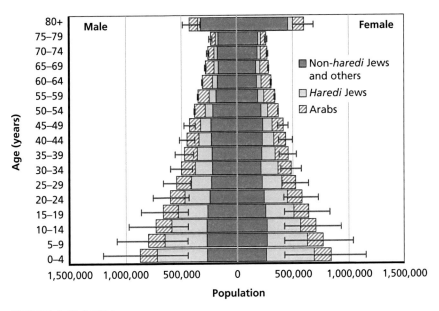

SOURCE: Paltiel, 2011.
NOTE: The blue whiskers flanking each bar show the range of possible projections within the fundamental medium-growth scenario assumptions.
RAND RR488-7.3

specialists has nearly doubled in the past ten years, Israel still has a relative shortage. General physicians, not geriatricians, fill most of the voids. Accordingly, hospitals might not have physicians who possess the appropriate skills to treat specific problems unique to the elderly population (Stessman, 2011). Also Israel has a relatively low nurse-per-population ratio. No formal academic specialization exists for geriatric nursing. However, some nurses are proficient in attending to the elderly and do take a central role in managing community care for the elderly. Further, although Israel has a government system supervising the care-

world: The average hospitalization period in Israel for the general population is 4.1 days and, for the 65-and-over age group, 5.5 days (Central Bureau of Statistics, 2010).

givers within eldercare facilities, many caregivers in the community still work outside the system and are therefore not supervised.

Housing and Living Environment

In Israel, the common approach is that the elderly remain in their homes and the responsibility for their care rests on family members. Most of the care for the elderly is thus provided by informal sources (Brodsky, Raznitsky, and Citron, undated). Eldercare outside the home and hospitalization is relatively rare (Habib, 2012). Therefore, a considerable and growing proportion of an adult's time might be devoted to caring for his or her elderly relatives.

The National Insurance Institute (NII) of Israel finances home care for elderly residents who have disabilities and remain in their homes according to their levels of functioning and activity as determined by NII functioning tests. Currently, 15 percent of the elderly in Israel receive subsidized assistance from around 80,000 caregivers. An elderly Israeli with severe disabilities is entitled to the assistance of a full-time caregiver (usually a foreign employee) 24 hours per day, financed in part by the elderly patient. NII expenditures grew significantly in the past few years and reached a record of 4 billion new Israeli shekels (NIS) in 2010 (Stessman, 2011).

Various services have been developed for elderly with functional impairment and low financial means, such as 170 activity centers, for the purpose of providing engagement, company, and hot meals. In 2008, 220 assisted-living facilities served more than 45,000 members. In addition, respite-care centers spread throughout the country offer the disabled elderly a time-limited stay, which also allows a break for the family members who are the primary caregivers (Central Bureau of Statistics, 2010, p. 29).

Despite this, many elderly in Israel report feelings of loneliness. Figure 7.4 provides a comparison of prevalence of loneliness among those ages 65 and older in Israel and that in several other OECD member countries as shown in results of a standard survey. This is important because of the growing body of literature demonstrating links between social isolation and loneliness on one hand and both

Figure 7.4
Share of the Elderly Reporting Feeling Lonely in the Four Weeks Prior to the Survey

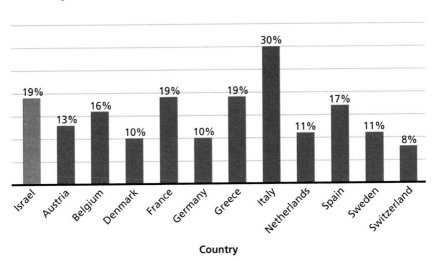

SOURCES: Survey of Health, Ageing and Retirement in Europe; Hebrew University of Jerusalem; Israel Gerontological Data Center, undated.
RAND *RR488-7.4*

mortality and morbidity on the other (Luo et al., 2012; Steptoe et al., 2013).

Most elderly prefer to remain in their homes and communities, often with the assistance of family members. However, in many cases, the families do not have sufficient training. Moreover, caregivers report a physical and mental load, as well as a noticeable deterioration in their quality of life and in their health, brought on by their responsibility for care (Brodsky, Raznitsky, and Citron, undated). An increased need to care for family members with disabilities could affect labor-force participation rates, as well as family members' hours of work. A need for additional manpower to care for the elderly could change the composition of employment in Israel; particularly, the rate at which people enter traditional and industrial sectors will diminish in favor of the service sector pertaining to the older population, which might, in turn, affect terms of trade (see below). There could also be even greater demand for foreign caregivers, accompanied by the possible difficulties such an increase implies.

Social and Long-Term Care Expenditures on the Elderly

Social expenditures on the elderly in Israel, such as pensions, social assistance programs, and transfer payments, are relatively low by OECD standards, representing some 5.5 percent of GDP (Shalev, Gal, and Azary-Viesel, 2012). This is nearly half the levels in other countries with larger elderly populations, such as Sweden (10.7 percent), the United Kingdom (10.4 percent), and Germany (9.7 percent). However, if one examines spending *per elderly person* in relation to per capita GDP, the gap appears smaller: Israel's 55 percent compares with Germany's 50 percent and the 65 and 67 percent, respectively, for Sweden and the United Kingdom. These figures reflect not only actual cost of services but also the degree of coverage and access.

Long-term home care is likely to prove increasingly significant with an aging population. Current costs for various categories of home care (community-based individual care, noncomplex institutional care, and complex hospitalization care) are NIS 11.4 billion (Chernichovsky and Regev, 2012). Currently, private and public sources of funding share these costs equally. This represents a relatively low share of GDP (approximately 1.2 percent), but, given the relative youth of Israel's population, this share is greater than the OECD trend when accounting for proportion of elderly in the general population. The proportion of such spending in Israel that comes from private sources is also high, exceeded only by Switzerland in comparison with 23 other OECD countries (OECD, 2011).

International Experience with Population Aging

How have other countries thought about population aging, and what have they done?

Israel will probably not mimic exactly what has occurred elsewhere. Any policy response must take into account the particular characteristics of population aging in Israel and the nature of its society and democratic processes. Yet, it is useful to consider the experience of countries already undergoing the changes now beginning in Israel. Not only is it possible to draw on their prior experience; examining other countries is yet another technique for shaping the conceptual frame of reference within Israel.

We were asked to examine as case studies Finland, France, Italy, Japan, Korea, Sweden, and the United Kingdom to determine what might represent standard approaches toward many of the same issues looming for Israel and what might be distinctive in terms of either analysis or policy. We provided detailed, country-specific results to the government socioeconomic strategic planning committee. Next, we provide an overview of those findings.

General Overview

Figure 7.5 shows Israel's relative position in the ratio of working-age population (ages 20 to 64) to those ages 65 and over. Although Israel has long been above the trend in most other countries and is likely to remain so, the differential between it and other developed economies will become considerably narrower. The relatively steady ratio of working to retired age groups that Israel has experienced since the early 1990s has begun a decline likely to continue for the next four decades.

Aging-related policies and planning are ongoing in all the countries examined in detail. There is considerable experimentation and policy modification even within a single country. Policies are formulated, are implemented (often starting in a pilot stage), and undergo revisions as needed. Policies generally go through adaptation while retaining the original policy formulation and the overall goals. This suggests the value of framing higher-level strategies to which policymakers can then refer when developing particular policies or in efforts of continuous improvement. Flexible adaptation becomes an asset to remaining on course if a larger strategic vision supports it.

A second main finding is that various issues that population aging affects (such as pensions, employment, housing, health, and family policy) appear most often to be treated separately and distinctly from each other. There has been some movement toward more comprehensive and integrated approaches in the United Kingdom, Korea, and France.

Figure 7.5
Ratio of Working-Age to 65-and-Older Populations in Selected Countries, 1950–2050

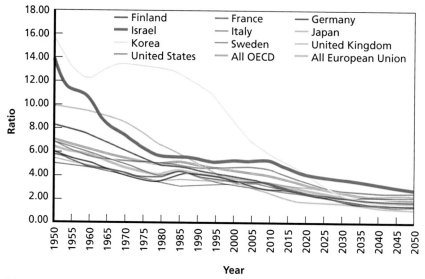

SOURCE: United Nations, 2008.
NOTE: The European Union consisted of 27 members at the time the United Nations compiled these data.
RAND RR488-7.5

For example, France's social cohesion plan of 2004 was an instance of an integrated approach to address a variety of social problems, such as unemployment, an increase in the number of socially excluded people, housing scarcity, disadvantaged regions, discrimination, and the education system. The plan also addressed several challenges related to the elderly. The plan was distinctive because, previously, each issue was addressed in isolation in a largely segmented approach (see, for example, Viprey, 2004).

The OECD has recognized population aging as one of the most daunting challenges that developed countries face. Within the European Union, the year 2012 was nominated as the European Year for Active Ageing and Solidarity Between Generations. The intention was to raise awareness of the contribution that older people make to society and to encourage policymakers and relevant stakeholders at all levels to take action with the aim of creating better opportunities for active aging and strengthening solidarity between generations. EU member states were encouraged to adopt aging-related policies. For example, the European Union issued a directive on discrimination (Council of the European Union, 2000) making age discrimination illegal in all member countries and giving three years to comply with this rule. Up until then, the Italian constitution prohibited any action that discriminated against workers according to their political or union affiliations, religious beliefs, races, or genders, but it did not explicitly mention age. The EU law caused Italy to add age to the list of forbidden bases for discrimination. Similar processes could hold sway in Israel, admission of which as an OECD member establishes various norms toward which the state would be expected to move.

We have arranged the following sections of this chapter topically to illustrate findings from the international comparison.

Health and Aging

Improvements in the health and functionality of the elderly (healthy aging) can enable them to live independently for as long as possible (Lafortune and Balestat, 2007). Healthy aging directly affects the costs of health and long-term care while increasing the well-being of the elderly. Thus, policies for healthy aging also play a role in mitigating future aging-related pressures on public finance. A healthy, older workforce could be less inclined to withdraw from the labor pool, under existing age-pension or disability arrangements.

A range of policies can affect healthy aging and can, when structured appropriately, interact and be self-reinforcing.[3] Because these can include socioeconomic, environment, and education policies, they often lie outside the normal scope of activities of health ministries. Policies for healthy aging are thus likely to require integration across a range of ministries. They can be grouped under four broad headings:

- **improved integration in the economy and into society:** Working longer provides an important social network. For those no

---

[3] We have drawn much of the treatment in this section from Oxley, 2009.

longer employed, healthy aging can also be promoted by better integration into society through participation in communal activities.

- **better lifestyles:** In this context, three areas of focus stand out: physical activity, nutrition, and substance use and misuse.
- **adapting health systems to the needs of the elderly:** Health care systems might be better adapted to the needs of the elderly through
  - more-regular follow-up of chronically ill patients and better coordination of care
  - enhanced preventive health services
  - greater attention to mental health
  - better self-care through increased health literacy and access to tools for greater adaptability.
- **attacking underlying social and environmental factors affecting healthy aging:** There are wide differences in health, morbidity, and mortality outcomes across socioeconomic groups. Age-adapted surroundings can encourage people to increase activity and can help break down social isolation and loneliness.

Social Integration

Healthy aging can be promoted by better integration of the elderly into society through participation in community and other social activities. It can lead to longer employment, which mitigates the financial burden on younger generations. Work is also an important social network in its own right. Longer employment has been demonstrated to be associated with better health (Alavinia and Burdorf, 2008; also see Parker-Pope, 2009). Finally, improved health status is usually correlated with higher quality of life and less strain on caregivers.

Care for People with Dementia and Alzheimer's Disease

The United Kingdom has adopted a dementia challenge, which sets forward the UK Department of Health's strategy (UK Department of Health, 2012). The key pillars of policy are to improve awareness of dementia, support early diagnosis, and improve care in all stages. Japan's dementia-related efforts have mostly been carried forward at the regional level seeking to develop supporters for dementia patients and to develop mutual assistance for elderly people with dementia, their family members, and social networks. Finland offers a memory-rehabilitation service through which social and health care professionals offer guidance and advice for people with memory disorders. Korea has two projects to test all senior citizens ages 60 and over for early signs and to mitigate deterioration from dementia at the earliest possible time.

Coordination of Care and Social Services

Sweden has launched a project to coordinate health care and social services for the sickest elderly. Sweden also works to streamline the use of resources to better address the needs of elderly people with the most-severe diseases. The Swedish government is developing mechanisms to reward results so that local solutions can emerge that are efficient, organized, and tailored to the conditions in each county and municipality. Efforts are currently under way to develop indicators and systems for outcome-based performance compensation.

Cost of Aging

Assumption-based projections foresee an increase in overall aging-related public spending, including pensions, health, and long-term care (Rechel et al., 2009). Yet, keeping people in good health and out of hospitals can mitigate overall health care costs.

Pensions, Income, and Welfare

Pensions and income-replacement policies are central in all the countries' strategies. Although it is a large and complex field in itself, several trends stand out from a cross-country assessment.

One indicator of retirement behavior is the average effective age at which older workers withdraw from the labor force, as shown in Figures 7.6 and 7.7.

In most OECD countries, the effective age of retirement is below the official age of eligibility to receive a full old-age pension. Japan, Korea, and Israel are notable exceptions. If nothing is done to promote better employment prospects for older workers, the increase in the number of retirees per worker will threaten living standards and

Figure 7.6
Effective Versus Official Retirement Ages for Men in the OECD

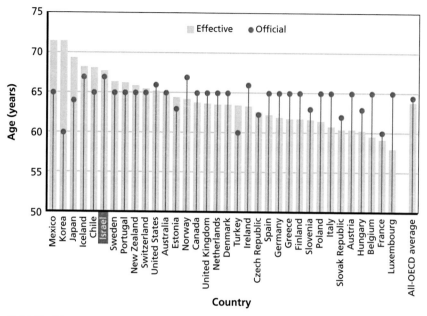

SOURCE: OECD, 2006.
RAND *RR488-7.6*

put enormous pressure on the financing of social protection systems (OECD, 2011).

In the past two decades, most of the countries have undertaken pension reforms. These reforms were motivated primarily by concerns about the financial sustainability of the pension system. Steps to reduce the cost of pension benefits promised for today's workers (in comparison with costs for previous generations) have included

- increasing the **eligibility age** for full pension benefits
- changing the manner of **calculating benefits**
- providing smaller **cost-of-living adjustments** and real pension increases than in the past.

Figure 7.7
Effective Versus Official Retirement Ages for Women in the OECD

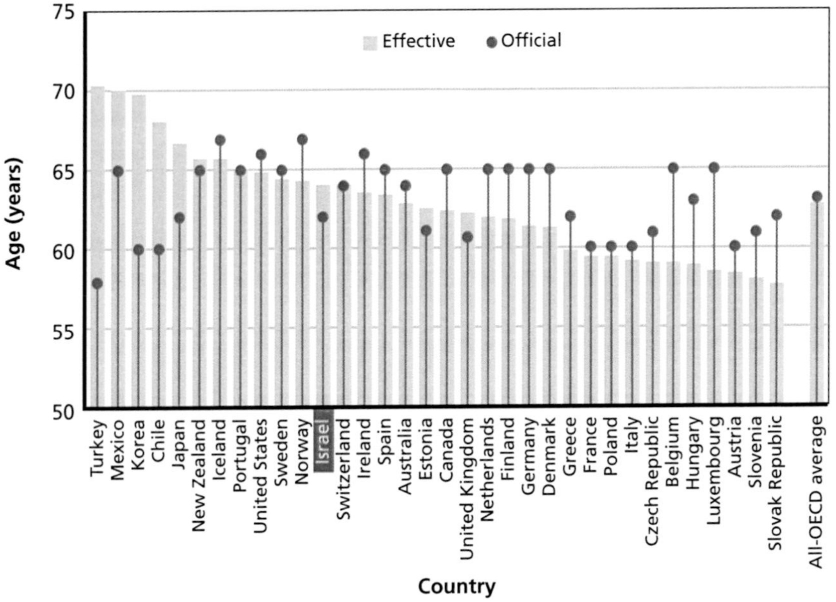

SOURCE: OECD, 2006.
RAND RR488-7.7

The OECD suggests three main options going forward for maintaining retirement-income adequacy without endangering financial sustainability (OECD, 2011):

- **longer working lives:** Action from government and employers on age discrimination, training opportunities for older workers, and working conditions needs to bolster pension reforms.
- **focused efforts:** Concentrate efforts of public retirement provision on the most vulnerable.
- **public education:** Encourage people to save for their own retirement to make up for reductions in public benefits that are already in the pipeline or are likely to be required.

The OECD's position is that a diversified pension system— mixing public and private provision and using pay as you go and prefunding as sources of finances—is not only the most realistic prospect but the best policy.

Although the complex structures and rules governing pension systems make these programs difficult to compare, our examination of pension reforms in other countries allows us to make several observations about recent trends.

Goals of Pension Programs

Pension programs generally have two main objectives: to redistribute income to low-income pensioners (thus preventing poverty in old age) and to help ensure that retirees have sufficient income to maintain an acceptable standard of living. Most countries pursue both goals in their overall pension policy, but emphasis varies.

Value of the Pension Benefit

Across OECD member countries, workers can expect their posttax pension to be worth an average of just less than 70 percent of their earnings after tax. Low-income workers in OECD countries will receive a net average replacement rate of about 85 percent. Some countries link contributions and benefits more closely. All have some form of safety net for older people, usually means-tested. The average minimum retirement benefit for full-career workers across OECD countries is worth a little less than 29 percent of average earnings (OECD, 2005).

Linkage to Other Issues

The overview revealed the interconnectedness between pensions and other key policy areas relevant to an aging population, including employment, health care, child and family policy, and financial planning:

- **tax rates, credits, and allowances:** Pensioners often do not make social security contributions; because personal income taxes are progressive, the average tax rate on pension income is typically less than the tax rate on one's usually higher former level of earned income. Older people in most countries also receive additional allowances or credits from the tax system.

- **consumer price indexing:** Nearly all OECD countries link pensions to changes in prices.
- **adjustment of past earnings:** Some countries adjust past earnings (known as valorization) to account for changes in the standard of living from when pension benefits are earned (i.e., when the person is employed and contributing to the system) and when benefits are claimed.
- **linkage to health policies:** By covering home care expenses and reimbursing pension recipients for extra costs caused by illness or disability, Finland enables pension recipients with illness or disability to live at home.
- **effect of life expectancy:** Differences in life expectancy have a large influence on pension wealth. Other things being equal, countries with lower-than-average life expectancy can afford to pay men a pension 10 percent higher than a country with average OECD mortality rates. In contrast, longer life expectancies necessitate lower pension levels (OECD, 2005).

Organization of Pension Systems

Many countries use two-tiered systems: a flat-rate scheme and an earning-related additional pension, both amenable to supplementation through voluntary private pensions. This characterizes the systems in the United Kingdom, Japan, and Finland. The systems used in Sweden and Italy more closely link benefits with actual contributions.

Employment and Retirement

Both Sweden and Italy emphasize flexibility in combining employment and pensions. These countries' systems allow people ages 61 and older to combine work and pensions. Japan also allows people to combine work and pension after the age of 65, provided that total income does not exceed a set level.

Benefit Levels and Retirement Age

In France, each additional year someone works after the pension age increases the benefit under the public scheme of the pension system by 5 percent. In Finland, the national pension can be deferred after age 65, and the pension is then increased by 0.6 percent for each month that retirement is postponed.

Withdrawal of Benefits

Upon retirement, Swedes and Italians choose how benefits are withdrawn. They may convert their pensions into fixed annuities to avoid investment risk. Alternatively, they may choose variable annuities in which funds continue to be invested by a fund manager of their choice.

Pension Benefits, Provision of Child Care, and Other Periods of Unemployment

Most of the countries reviewed ensure that pension contributions continue even when parents are engaged in child care or are unemployed.

Employment and Work Conditions

The key message from the review of country-specific policies is that, if there is no change in work and retirement patterns, the ratio of older inactive persons to employed workers will almost double from around 38 percent in the OECD in 2000 to more than 70 percent in 2050 (Schnalzenberger and Winter-Ebmer, 2008). Furthermore, currently inactive older people are a potential source of additional labor. Employment and social policies and practices that discourage work at an older age despite longer and healthier lives result in the loss of a valuable resource (OECD, 2006).[4] Conversely, continued employment is associated with better health status among the elderly. In this regard, blue-collar and less-qualified workers are more likely to retire earlier than white-collar and more–highly qualified workers.

If working longer is to be an attractive and rewarding proposition for older people, there is need for action on both the demand side and supply side. This requires cooperation by government, employers, and civil society. First, there must be strong financial incentives to continue

---

[4]   One could argue that policies to encourage greater immigration, higher fertility, or faster productivity growth could offset these negative consequences. Although these developments could all help offset negative effects in some countries, they are less relevant to Israel, where immigration forecasts are not high and the fertility rate of 3.0 children per woman is still well above the replacement level.

working, and subsidized pathways to early retirement would need to be eliminated. Second, employment practices of firms would need to be adapted to ensure that employers have stronger incentives to both hire and retain older workers. Third, older workers require appropriate accommodation and encouragement to improve their employability. Finally, both employers and older workers will need a major shift in attitudes about working at an older age (OECD, 2006).

There is a widespread concern that more employment among older workers might come at the expense of younger workers. International experience shows that employment trends are similar for all age groups: Countries that have managed to raise employment rates among older people have done so in a context of higher overall employment and falling unemployment. Research suggests that policies of squeezing older workers out of the labor market have not created more job opportunities for other age groups (Gruber, Milligan, and Wise, 2009).

Disincentives for Early Retirement

Early exit from the labor market tends to be a one-way process, with very few older workers returning to employment. One of the reasons for this irreversibility is employers' reluctance to hire older workers.

Challenges for the Elderly in Maintaining Employment

Employers might doubt the ability of older workers to adapt to technological change. Employers are also concerned about wages and nonwage labor costs that rise more steeply with age. To counter negative employer attitudes, several countries have introduced age-discrimination legislation or information campaigns. Some have also been taking action to address the factors that prevent employers from hiring and retaining older workers, including wage subsidies (OECD, 2006).

Changes in Expectations on Continued Work

Older workers might wish to continue working, albeit with less intensity, or to receive accommodation, such as being allowed to telecommute. Some countries have begun to adapt working conditions for older workers, e.g., by facilitating access to part-time jobs and developing flexible work arrangements (OECD, 2006). To prepare older people to stay in the changing workplace, such countries as Finland have been cultivating a culture of lifelong learning at midcareer, while others, such as Japan, have been providing employment assistance to older people. Another approach is to give special incentives for private employment agencies to place older job seekers in jobs. However, one issue not addressed in the policies reviewed is the need to help older workers prepare for greater job mobility at the end of their careers.

Education

Education is deeply connected to many of the issues raised by an aging population. We highlight three of these.

Education and Employment of the Elderly

Education and retraining can maintain longer work life for current and future workers. There is also a strong correlation between years of formal education and both longevity and quality of life in older years (although there is as yet no definitive proof as to which direction the causation might go; see Montez, Hummer, and Hayward, 2012). But education is important to help older people obtain and retain good jobs and associated rewards, especially in an information-based economy.

Education and Personal Development

Educational opportunities can provide seniors with information to support a healthy lifestyle and provide access to valuable networks and social relationships. Education can help seniors sharpen their cognitive skills and reinforce their sense of control over their environment. Some have argued that the connection between continued mental and physical activity is key to maintaining health in general and forestalling cognitive deterioration, although further research is needed to support this point. One example of direct linkage at the policy level involves Finland's support for continuing education. The Universities Act of 2009 states that, "in carrying out their mission, the universities must promote lifelong learning" (Finnish Parliament, 2009, Chapter 1, § 2, ¶ 1). In the United Kingdom, several initiatives have had the objective of reducing the obstacles for older people to have access to the Internet. There have also been explorations of outcomes from intergenerational educational experiences (Springate, Atkinson, and Martin, 2008).

Ensuring a Sufficient Number of Health Care Professionals

The educational system produces the skills required in care- and health-giving professions and the workers required to replace the skills of those passing into the retired population.

Educating the Young and Productivity

There is another potential connection between education and aging from the standpoint of national policy. To the extent that resources to educate the younger generation and resources to support an aging population both draw from public funds, there is potential for inter-generational conflict. However, as fertility declines, per-family and per-child expenditures on human-capital development rise in a surprisingly consistent manner across countries (Lee and Mason, 2010). The effectiveness of these expenditures has at least the potential to raise the productivity of succeeding generations sufficiently to partially counteract the effects of increasing elderly dependency ratios.

Using Seniors as Resources for Education

In the United States, the Experience Corps has been run as an experiment in the Baltimore area. It involves bringing in seniors as classroom aides and enrichment resource providers and involving them directly in the educational experience. The initial results show not only positive outcomes in seniors' health and life quality but also improved educational results in the classrooms in which they have been deployed (Martinez et al., 2006).

Housing

Housing relates to the challenge of population aging both directly and indirectly in such areas as health, transportation, and quality of life.

Effect on Daily Functioning

Housing conditions can affect a chronically ill or disabled elderly person's ability to function in performing everyday activities. The ability to carry out such activities can mean the difference between an independent old age and one that requires institutionalization. The nature of the housing available to an elderly person will affect whether he or she must transition from minimal assistance requirements into housing arrangements requiring greater public assistance.

Effect on Caregiver Services and Support for Aging in Place

Housing can affect the ability of care providers from outside of the household to render services. In addition, housing characteristics might affect a family's willingness to care for an elderly relative (Struyk and Katsura, 1988). Investments in communities to facilitate "aging in place" or support for caregivers can often promote social integration (Cannuscio, Block, and Kawachi, 2003).

Access to Social Capital

Social connections enable seniors to maintain productive, independent, and fulfilling lives and so are an ingredient of successful aging. Housing circumstances can determine who can remain in the community. The ability to tailor housing options might also affect how family intergenerational wealth is distributed.

Transportation

Eventually, almost all older people will need access to alternative transport modes. Policies and programs related to transportation address the following issues.

Safety

Older drivers tend to be safe drivers. Nevertheless, there is concern about declines in reflexes, ability to respond, and decline in driving skills at older ages, as well as consequent vulnerability to personal injury in a crash. Additionally, older pedestrians have higher fatality rates than younger ones. Thus, the OECD has developed recommendations for building safer vehicles for the elderly and investing in safety-enhancing road infrastructure.

Environmental Issues and Health

Population aging exacerbates the effects of societal transportation problems, such as air pollution, congestion, and environmental degradation, because seniors are more vulnerable to health issues related to these phenomena (Cobb and Coughlin, 2004).

Lifelong Mobility

Given that a growing number of seniors tend to remain in their own homes, well-planned transportation systems can also promote lifelong mobility (OECD, 2001a).

Family Policies

Credible research on the effect of various family and immigration policies in offsetting population aging is scarce. For each example of policy impact, there is a counterexample with no impact. New policies are usually introduced in bundles, so individual policies cannot be evaluated independently from each other.

Nonetheless, developed countries clearly consider family-related policies that encourage higher fertility to be useful tools to address the challenges associated with aging societies. No country assumes that family policies alone can offset aging's negative effects; where implemented, such policies usually accompany broader attempts to mobilize all available labor reserves in order to sustain economic growth.[5]

Family policies are interwoven into employment, pension, health, and other policy areas. Policies regarding taxation, social welfare benefits, work leaves, pensions, and other areas all consider one's family status and whether one has children. Explicit policies attempt to encourage young couples to have children either by making it easier— and more affordable—to raise children or by offering specific inducements to have more children.

Understanding the System: Mapping onto Socioeconomic Outcome Indicators

After these initial examinations of available information on aging, we can summarize several important themes by turning to the framework for socioeconomic outcomes and measures developed in Chapter Four

---

[5]   A possible technological dimension might come to play a larger role. See, for example, Hoorens et al., 2007.

to illustrate how population aging and its associated phenomena fit within a more general perspective on future socioeconomic well-being.

When we initially presented this framework, it served as a means for characterizing positive or negative socioeconomic futures for Israel. Now, we reverse the flow. Having conducted assessments as described in this chapter and in those to follow, we select in Table 7.1 from the

Table 7.1
Dimensions and Metrics of Socioeconomic Outcomes Affected by Population Aging

Dimension	Ideal Metric	Candidate Measure for Indicator
Socioeconomic well-being of individuals and families	Availability of material means to achieve life goals	Household net adjusted disposable income per person
	Wealth to sustain shocks and meet goals over time	Household financial net worth per person
	Access to adequate housing	Number of rooms per person
	Housing affordability	Housing costs as share of adjusted disposable income
	Status with respect to disease	Life expectancy at birth
		Self-reported health status
	Status with respect to wellness	Self-reported limitations on activity
		Incidence of dementia and cognitive deterioration
		Rates of overweight and obesity
	Time balance between paid work, time with family, commuting, leisure, and personal care	Employees working very long hours
		Time devoted to leisure and personal care
	Capacity for and availability of lifelong learning	Share of those ages 25 and older who have engaged in further education in life skills, work skills, and general culture

Table 7.1—Continued

Dimension	Ideal Metric	Candidate Measure for Indicator
Socioeconomic well-being of individuals and families, continued	Social network support	Availability of help through social contacts and families
	Frequency of social contact	Meet with social contacts and family at least once per week
	Water and air quality	PM_{10} concentration
	Fear of crime	Self-reported victimization
	People's overall views of their own lives	Life satisfaction
	People's present sense of satisfaction	Affect balance
Sustainable, growing, and innovative economy	Domestic	Labor productivity
	Fiscal and monetary balance	Share of primary civilian expenditures in total government spending
Democracy, effective government, and access to opportunity	People's perceptions of the adequacy of services	Waiting times for service, e.g., transfer from emergency room to hospital bed

NOTE: PM_{10} = particulate matter up to 10 microns in size.

full list of measures (and candidate indicators for policy) those that seem most affected by the aging of Israel's population. This renders the full matrix of socioeconomic indicators both a tool for analysis and a preliminary framework for active strategic and operational policy planning.

In this case, comparing the trends and factors affecting population aging in Israel and the matrix we developed for understanding socioeconomic outcomes in Israel affirms the aging issue as an important area of focus as well from the perspective of long-term socioeconomic strategies for the nation. Similar analyses for other issues done in this format can assist in conveying a wider understanding of the linkages in contributing to socioeconomic outcomes. This approach allows systems of interest to be mapped individually, in relation to one another, and from the perspective of the national interest.

CHAPTER EIGHT

Dynamic Analysis: Focus on Health Care Cost and Dependency

In Chapter Seven, we assessed available information and drew infer-ences for Israel. In this chapter, we utilize existing data on trends to conduct *dynamic analyses* projecting what future socioeconomic out-comes might be under various assumptions about the continuation of those trends. In particular, we look at implications for health care costs and for several factors that will, in part, determine future GDP for Israel.

Changes in Health Care Costs

The aging of Israel's population will affect health care costs as the elderly consume more health services than either teenagers or working-age adults do. We present two analyses of trends in health care costs. The first focuses on international data and trends and the second on specific trends in Israel.

A widely cited study examined age-specific relative health care costs across ten OECD countries, as presented in Table 8.1.[1] We nor-malized the relative costs by age category to the share consumed by those ages 50 to 64.

---

[1] We based all calculations on data found in Hagist and Kotlikoff, 2005. Israel was not among the countries Hagist and Kotlikoff examined, and, to the best of their knowledge, a similar analysis had not been applied to Israel. Hence, in this report, we have used their data as a proxy for age-specific differences in health care consumption in Israel.

Table 8.1
Relative Shares of Health Care Consumption, by Age Group

Age Group, in Years	Relative Share of Health Care Consumption[a]
0–19[b]	0.55
20–49	0.59
50–64	1.00
65–69	2.06
70–74	2.29
75–79	3.45
80+	4.36

[a] We have normalized the relative shares to the share for the 50- to 64-year-old age group.

[b] For this table, we combined the relative shares of the zero- to four-year-old and five- to 19-year-old age groups and weighted them by years.

Table 8.1 shows that health care for someone age 80 or over costs more than four times what it costs for someone in the 50- to 64-year-old reference group. As the average age of Israel's population grows, there would be an effect on total health care costs just from the shifting of shares among age groups, even if nothing else were changed.[2]

In 2009, Israel spent approximately 7.9 percent of its GDP on health care. A little less than 60 percent of this total was public spending, with the rest being private. Table 8.2 shows how health costs would increase as a share of GDP if the nation maintained the same level of GDP and population but we altered the shares of population to conform with those that several of the CBS demographic scenarios for 2059 presented.

Population aging alone (with no other related effects) would require Israel and Israelis to spend at least 15 percent more on health

---

[2] There are, of course, good reasons to believe that other changes might also occur in terms of both GDP growth and the prices of various types of health care. Both would most likely be in directions that would further accentuate the rise in health care spending as a share of GDP.

Table 8.2
Health Care Costs for 2009 If 2009 Population Shares Resembled Those for 2059 Under Different Demographic Assumptions

Demographic Scenario	Health Care Share of GDP	Per Capita Cost (2009 NIS)	Percentage Change Relative to Actual 2009 Health Care Costs
2009 actual	7.9	8,071	0
Age shares as in 2059 high growth	9.0	9,243	15
Age shares as in 2059 medium growth	9.8	9,980	24
Age shares as in 2059 low growth	10.9	11,118	38

care in 2059 than they actually did in 2009, even under the most favorable of the three CBS population scenarios. These costs would increase dramatically if the low-growth CBS scenarios came to pass. This analysis does not take into account further effects caused by having a smaller working-age population with possible effects on size of GDP.

The effects of aging are also apparent for projections to 2029, as shown in Table 8.3. The projections for 2029 vary much less than those for 30 years later. Yet, health care costs will increase at least 10 percent

Table 8.3
Health Care Costs for 2009 If 2009 Population Shares Resembled Those for 2029 Under Different Demographic Assumptions

Demographic Scenario	Health Care Share of GDP	Per Capita Cost (2009 NIS)	Percentage Change Relative to Actual 2009 Health Care Costs
2009 actual	7.9	8,071	0
Age shares as in 2029 high growth	8.7	8,895	10
Age shares as in 2029 medium growth	8.8	8,974	11
Age shares as in 2029 low growth	8.9	9,076	12

solely because of the shift in age distributions. Further, this change will occur over a relatively short time; indeed, it has already begun.

A similar analysis focuses specifically on care for the elderly. In 2010, the national expenditure (both public and private) on elderly health and long-term care (ages 65 and over) was around NIS 18,000 per capita. Seventy percent of the total health expenditure is publicly financed. Figure 8.1 shows a forecast that total spending from public and private sources on elderly health and long-term care in both community centers and hospitals would claim a 36-percent larger share of GDP in 2030 than it did in 2006.

The distinction between public and private finance is not trivial. Even quite wealthy countries can find such costs straining government finances. The fact that a national economy generates income does not mean that the government can command resources to meet its obli-

Figure 8.1
Projection of Public Expenditure for Health on Elderly (65 and over) as Share of GDP

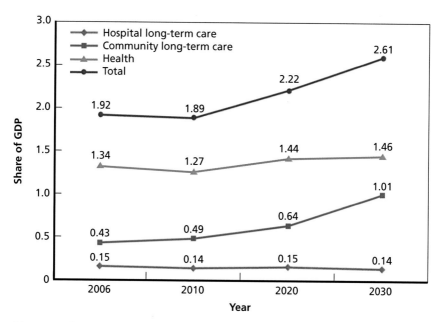

SOURCE: Habib, 2012.

RAND RR488-8.1

gations. Government attempts to raise revenues through higher taxes come at a cost to overall economic efficiency. Government obligations for social insurance and public health insurance schemes matter.

Changes in GDP

Dependency Ratios

The previous section noted the changes to health care costs as a result of shifting age distributions. Another, related issue is the share of the population that will be available to pay for these and other costs. The dependency ratio compares the number of working-age adults and the number who are either above the age of pension or too young to work.

In Israel in 2009, this simple dependency ratio was 0.84. That is, for every 100 Israelis of working age (between 20 and 64 years of age),[3] 84 people were either younger than 20 or at least age 65. In other words, each working-age person was responsible, on average across the entire population, for the support of 0.84 non–working-age dependents.[4] It is also possible to calculate the old-age dependency ratio by leaving out the youth population. In 2009, for every 100 working-age Israelis, only 18 were beyond the age of 65. This is a relatively low burden when compared with those in other countries at Israel's level of economic development. Up until now, the age distribution in Israel resembled those in developing countries in being skewed toward a younger average-age population.

This favorable historical circumstance is about to change dramatically. By the year 2024, the old-age dependency ratio will have increased

---

[3] The standard often used for international comparison is to designate the working-age population as ages 15 to 64. This becomes less useful as a definition for developed countries, in which work is often postponed for higher education and even less so in Israel with its general military conscription at age 18.

[4] Although this measure is regularly used to assess dependency burdens, it is, of course, a crude measure. Not all those who are of working age actually produce outputs measured in standard national income accounts. Not all those ages 65 and over have ceased to earn an income or require drawing to the full extent on the resources of others, whether public or private, to sustain themselves. But this measure serves to provide comparisons across time and is a benchmark against other countries.

by 35 to 46 percent. Depending on which CBS scenario one selects, 100 working-age Israelis will be supporting an average of 24 to 26 aged dependents. Although calculations beyond that point become less certain, the trend will not reverse itself in any of a wide range of scenarios. In another ten years, 2034, the range rises to between 27 and 29. In 2059, the high–population growth projections yield 31, while the low-growth projections, those that seem to accord with more-recent trends in fertility and age expectancy, yield 38 people above age 65 for every 100 of working age. Under the high-growth scenario, on the other hand, the large share of young would create a squeeze from the other end of the age distribution. That scenario would yield a *total* dependency ratio of 1.24, compared with 0.84 in 2009. Indeed, under those assumptions, the crossover point—that is, when more dependents are in the population than workers—could come relatively soon—in the year 2020 or early 2021.

Figure 8.2 allows comparison between Israel and several other OECD countries, as well as the average of all member states of the OECD, for the age-dependency ratio. It demonstrates the benefit that Israel receives from its younger age structure. But it also shows that a long period of a relatively fixed age-dependency ratio has come to an end. Israel will remain comparatively well off compared with the other OECD countries, but, not long after the year 2020, Israel's age-dependency ratio will be similar to the current average among OECD countries. This level has led the OECD to call for a specific focus on the issues raised by the aging of societies within its member states (OECD, 2006).

Effects of Labor Participation

For reasons we have discussed, the simple dependency ratios we used previously certainly do not accurately portray the true facts of dependency and burden in any country.[5] Nevertheless, if one can have some

---

[5] Panel on a Research Agenda and New Data for an Aging World et al., 2001, pp. 42–43, examines several alternative dependency ratios that adjust the numerator (those ages 65 and older), the denominator (those in the working-age groups), or both to deal with the issues of economically active seniors on the one hand and economically inactive younger people on the other. When applied to data on various countries at different stages of development,

Figure 8.2
Ratio of Elder (65 and over) to Working-Age (20 to 64 Years of Age) Cohorts in Selected OECD Countries, 1950–2050

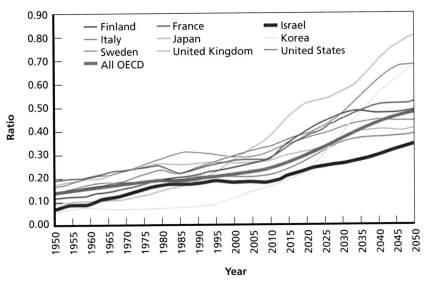

SOURCE: United Nations, 2008.

RAND *RR488-8.2*

confidence in broad assumptions about similarities across time and across countries, they serve a useful purpose. In Israel, this becomes problematic.

There is a structural component to Israel's labor-force participation that could exacerbate the effects of these demographic trends. The figures on labor participation show that two important segments of Israel's working-age population, *haredi* men and women from the Arab community, have a lower rate of labor participation than the rest of the population does. According to the OECD, 76 percent of the non-*haredi* population of Jews and non-Arabs were active in the labor force

---

a dependency ratio of the number of economically inactive older people to the total pool of those economically active regardless of age is generally higher than the standard measure in developed economies (with the exception of Japan, where there is a tradition of continuing working life). It was slightly lower than the standard simple measure in the developing economies of Mexico and Japan (c. 1997).

in 2009. This compares to only 39 percent for *haredi* men and 25 percent for Arab women.[6] This changes the perspective on the relatively favorable current dependency ratio in Israel and what the implications might be for the future.

For this reason, we constructed a more detailed, but still simplistic, approach to better understand the consequences of this structural element in Israel's economy and society in the presence of an aging population. Instead of just using the number of people in the 20- to 64-year-old age groups as the denominator and proxy for the working population, we now weight male and female Arabs, *haredim*, and others by their labor-force participation rates in 2009.

This new labor-force participation–adjusted age-dependency ratio was 0.26 in 2009, rather than the previously calculated ratio of 0.18. Similarly, the new adjusted total dependency ratio was not 0.84 but rather 1.21. That is, the number of actual dependent Israelis ages zero to 19 and at least 65 years old exceeded the number of Israelis actually employed in the formal economy. If we assume that the labor-force participation rates of 2009 continue to hold true in the coming years, the new, labor-adjusted total dependency ratio could increase from 2009's 1.21 to 1.99 by 2024.

If nothing else were to change, the only way to keep the increase in the labor-adjusted dependency ratios in line with those for simple dependency would be for increased labor-force participation from the *haredi* community and by Arab women. Of course, the level of such increases and when they might occur would both be crucial in determining the size of this offsetting effect.

We can examine this by supposing changes in policies that would lead to a 1-percent annual increase in the share of the members of the *haredi* and Arab female working-age populations that do, in fact, enter into the work force in each year between 2009 and 2024. We further assume that they hold at the 2024 rate in subsequent years. In other words, over that time span in each of those groups, the number actively

---

[6]   Arab-sector men were at the 71-percent level while the rate for *haredi* women, although higher than the rate for men in that community, had a labor-force participation rate of only 58 percent (Ishi, 2012).

engaged in the labor force would increase a nominal 15 percentage points. For Arab women, for example, this would mean a change from a labor-force participation rate of 25 percent in 2009 to 40 percent in 2004. This assumption, therefore, would translate into a 60-percent increase in the labor-force participation rate for that particular group— a fairly dramatic change in such a short time period.

If both groups currently less engaged in the labor force were to experience a uniform 15-percent increase, the growth rate of the labor-adjusted total dependency and age-dependency ratios would then closely match those of the traditional ratios in the years 2024 to 2034. This rough equivalency would hold in the later years as well, under the high-growth scenarios. The lower-growth scenarios would have higher growth in labor-adjusted dependency ratios in the later-year period.

Macroeconomic Effects of Population Aging

Labor-force participation rates are just one of several possible effects that could have macroeconomic consequences for Israel. For example, the pattern of consumption for older people is different from that for people who are younger or are in the early or middle stages of their working lives. In a country the size of Israel, this could affect the structure of manufacturing and trade. Some might point to the pattern of innovation, creativity, and productivity and suggest that a labor force more skewed toward the older end of the spectrum might well prove less innovative or productive in aggregate. There might also be effects on exchange rates as the balance of demand shifts from tradable toward nontradable goods and services (Braude, 2000).

The U.S. National Research Council examined the available evidence on these and similarly prospective macroeconomic effects stemming from population aging (Committee on the Long-Run Macroeconomic Effects of the Aging U.S. Population et al., 2012). The overall message from this metastudy was twofold. First, the committee found that *for the United States*, such phenomena as shift in types of assets purchased, aging-related changes in productivity, differing patterns of consumption, and effect on living standards are likely to be small (although a good deal of uncertainty remains about actual magnitudes or even direction). This notwithstanding, the second message is that the best way to ensure against unfavorable developments or surprises is to take measures for

> sensible policies [to be] implemented with enough lead time to allow companies and households to respond. The ultimate national response will likely involve some combination of major structural changes . . . , higher savings rates during working years, and longer working lives. The committee called attention to the cost of delaying our response to population aging. The longer our nation delays making changes to the benefit and tax structures associated with entitlement programs for older individuals, the larger will be the "legacy liability" that will be passed to future generations. The larger this liability, the larger the increase in taxes on future generations of workers, or the reduction in benefits for future generations of retirees, that will be required to restore fiscal balance. Decisions must be made now on how to craft a balanced response. (Committee on the Long-Run Macroeconomic Effects of the Aging U.S. Population et al., 2012, p. 4)

These findings were predicated solely on the conditions and experience of the United States. Strategists in Israel might find these conclusions and the panel's reasoning to be of value in shaping their thinking about Israel's future. They cannot substitute for them.

Social Security

Some social insurance payments in Israel are conducted under the pay-as-you-go system and some through defined-contribution accumulation. The share of pensions in GDP is stable at around 6 percent (NII, 2011). Most is financed by workers' contributions, with only a small part being financed by the government.

The amount held in reserve to meet pension obligations has increased in recent years. A strategic committee of the NII chaired by Yehuda Kahane recommended a reserve ratio adequate to cover 2.5 years of payments, even without further proceeds (NII, 2011). With business as usual, the reserve ratio is expected to change its course and begin decreasing, starting from 2014. According to demographic forecasts and with no change in policy, in 2031, it could drop below 2.5 years and around 2050 would reach zero.

The Real Exchange Rate and Prices

There is a possible mechanism through which differential rates of population aging could affect domestic prices and exchange rates (Braude, 2000). To the extent that the so-called law of one price holds, the ratio of the nontradable goods in each country would determine the real exchange rate between two currencies. An increase in the rate of the unemployed or those who demand more services (the classic nontradable good) would raise the price of nontradable goods. This could have relevance in the following ways if the share of elderly in the population increases:

- Both the elderly and the young consume more nontradable goods (e.g., education and health care) than tradable goods; therefore, an increase in the percentage of these age groups in the population will increase the demand for nontradable goods.
- Both the elderly and the young tend to save less and consume more (of all products and goods relative to their life-cycle average savings and consumption), which also increases the demand for nontradable over tradable (capital) goods.
- The elderly have greater purchasing power than their income, which is derived from their prior savings. They therefore add purchasing power without adding to the labor supply (ignoring potentially sticky saving habits), which increases the demand for labor and therefore increases the demand for nontradable goods.

The price of tradable goods is derived from international trade, and local demand does not affect it (because Israel is a small and open market and prices are set in the global market). The price of nontradable goods, however, can change; an increase in the share of the elderly population will create an increase in the prices of nontradable goods, and this, in turn, will increase the real exchange rate.

Foreign Direct Investment

A focus on GDP—that is, the value of goods and services produced within a country—should not obscure the potential importance of a nation's earnings that could stem from a positive foreign investment position. The means for financing the retired cohort's consumption does not all have to be produced domestically. This factor can be particularly important for a small, open economy, such as Israel's, which has experienced large swings in net foreign direct investment in recent years. For this reason, any strategic perspective on aging in Israel and the implications for both dependency ratios and gross national product should take this factor into account.

CHAPTER NINE

Scenario Building: First Simple Steps

In Chapter Eight, we reported on *dynamic analyses* to project current and possible trends into the future. The quantitative approach is indispensable in effectively developing a strategic perspective in the available data. But to the extent that thinking strategically also requires recourse to the process of *inductive*, as well as deductive, reasoning—that is, a systematic consideration of the question "what if?" in its various forms—more-qualitative approaches can add to the projections that derive from more-analytical methods. In this chapter, we use two approaches to *scenario building*, each deliberately chosen for its relative ease of implementation.

By its nature, strategy requires thinking in time by imagining future worlds, either attractive or undesirable, and outlining courses of action that will either achieve or avoid them. This requires an ability to imagine alternative futures. Different tools have been designed to enhance the ability to think effectively about the future.[1] The task is a challenging one. We must overcome the "tyranny of the present": all of the sensory and cognitive richness of the current world that makes it difficult to imagine fundamentally different futures or even those changed only in limited ways. Any particular technique for scenario thinking is successful only if it can break that hold. Some methods might be better suited to some purposes than others. They also vary in the investment required, the difficulty of their application, and the degree to which they are accessible to wide audiences.

---

[1] A rich library of such techniques can be found in Glenn and Gordon, 2009.

Although a method might prove valuable, none can be more than ancillary support to the process of group reasoning. A method should never be allowed to become the focus of the scenario exercise. The goal is to provide an occasion for people with differing perspectives, backgrounds, and interests to gather willingly and contribute to narratives that convey meaning about possible futures.

We illustrate this point with one of the simplest techniques, the futures wheel. Its framework is shown in Figure 9.1. At the center is some prospective change that might arise from extrapolation of cur-

Figure 9.1
Basic Form of the Futures-Wheel Framework for Scenario Building

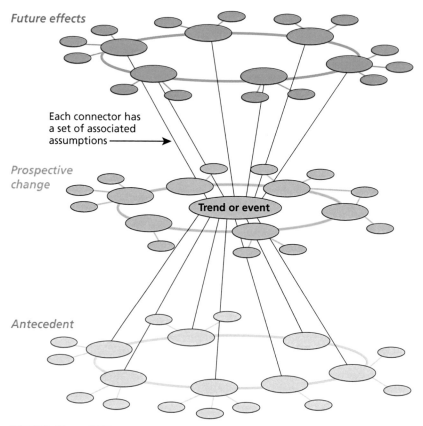

SOURCE: Glenn, 2009.
RAND *RR488-9.1*

rent trends. The resulting prospective change, also possibly affected by or connected to other contemporary factors, can, in turn, have consequences further into the future.

Figure 9.2 shows use of a futures wheel as an aid to extrapolation. It notes at least three trends observable in 2012: increasing longevity, increasing aged population, and the aged generally requiring more health and social services, especially those focused on problems of aging, than people in their working years do. For the sake of simplicity, we do not show collateral issues, such as budgeting constraints, politi-

Figure 9.2
The Futures Wheel Used for Event Extrapolation

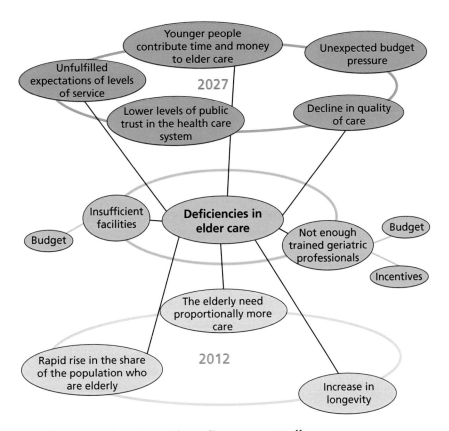

SOURCE: Authors' discussions with Israeli government staff.

RAND RR488-9.2

cal constraints, and other contributory factors. The result shown in the central portion is a foreseeable possibility of shortfalls in elder care as demand outstrips the supply.

The upper tier—the future effects—displays some possible consequences of supply deficiencies. These include unexpected budgetary pressures, decline in service and standards of care, the young needing to care for older family members financially and practically, failure of the level of care to meet public expectations, and a possible general loss of public trust in the health care system. We provide this narrative solely in qualitative terms. It provides an entry point for the use of other techniques geared toward quantitative assessment.

An alternative approach to the futures wheel might start with some prospective change that policymakers might desire or that might, alternatively, prove to be a challenge to achieving policy goals. In Figure 9.3, the analyst places at the center a medium-term goal that policymakers would like to achieve or a condition they want to see in place, and then reasons backward. Our prior analyses suggest that one reasonable strategy for dealing with aging populations is for people to remain active as workers. The analyst then considers what would be necessary for this to take place. Some factors might be financial incentives to firms for hiring or retaining older workers; incentives to the workers themselves (which might include, for example, liberalizing rules of pension eligibility to allow for employment); conditions of work, including access to retraining and elimination of discriminatory practices; and a change in public attitudes toward work and retirement. The upper tier then shows where one might expect long-term effects from the prospective change in areas, such as intergenerational transfers, effects on social and family life, the balance of supply and demand in the labor market, and the volume and categories of public expenditures.

The futures wheel, though simple in nature, provides a vehicle for structuring a discussion of different possible futures and their antecedents, as well as for framing the results of such a discussion.

Figure 9.3
The Futures Wheel Used for Policy Exploration

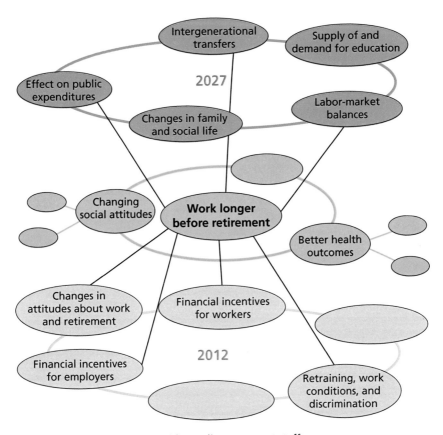

SOURCE: Authors' discussions with Israeli government staff.
RAND RR488-9.3

"In the Absence of Action, What Could Go Wrong?"

Using such techniques as the futures wheel in discussions with Israeli government staff, we identified problems that might arise as a basis for asking further questions. Here are the principal issues we identified using this process, possible negative outcomes one might see as a result of the population aging, absent any preparation:

- reduced housing turnover and a mismatch between supply and need
- increased occupancy rates in health facilities
- growing share of healthy, idle retired
- no provision made for use of retirees as aides, teachers, substitutes, tutors, or resources
- pressure on education and training budget limiting general skill improvements
- insufficient training in geriatric medical, therapeutic, and assisting professions to meet needs
- fiscal pressures reducing education funding
- skill training for later life minimized; little older-age enrichment
- insufficient professional staff and facilities for elder health care
- dementia and cognitive deterioration becoming increasingly prevalent
- even if they are efficient, government services failing to meet the expectations of a larger elder population
- quality of health service, e.g., waiting times
- limited retraining and workplace accommodation limiting contributions from the able retired
- loss of key functions as the age cohort retires without replacement or ready trainees
- possible effects on productivity as the workforce ages and changes composition
- larger share of older voters who vote their economic interests
- delayed or reduced generational transfer of wealth
- continued traditional approaches failing to create a path for second careers
- loneliness and lack of engagement affecting health
- loneliness and depression being pervasive
- the young contributing more time and money to elder care
- elders demanding more public and alternative transportation
- pay-as-you-go systems needing more workers' taxes
- elder-care burden affecting family budgets
- financial literacy being low: inadequate savings and poor asset allocation

- increased demand and ready supply increasing share of elder-care workers who come from other countries
- capital flowing to "younger" countries with higher returns on investment
- change in balance between tradable and nontradable production affecting the exchange rate.

One of the values in generating a list of this type is to allow for greater attention to be focused on particulars while gaining a better understanding of how factors, outcomes, causes, and effects, might be related. In Table 9.1, we take several items from this list and, for each, enumerate some of the driving factors that would lead to these negative outcomes.

Table 9.1
Population-Aging Factors Driving Selected Outcomes in the Absence of Preparation

Possible Negative Outcome	Driving Factor
Growing share of healthy, idle retired	Declining share of working-age population; extended retirement years; regulations on pension and retirement; regulations on employment; few models of work–life balance; insufficient focus on retraining; employer and workplace attitudes; social patterns excluding elders
Insufficient training in geriatric medical, therapeutic, and assisting professions to meet needs	Insufficient incentives for geriatric-worker training; regulations on employment; regulations on health care; insufficient training in self-sufficiency
Insufficient professional staff and facilities for elder health care	Increased costs of elder care; competing public social needs; regulations on employment; insufficient training in self-sufficiency; regulations on health care
Cognitive deterioration becoming increasingly prevalent	Extended retirement years; regulations on pension and retirement; regulations on employment; social patterns excluding elders; few models of work–life balance; insufficient elder-enrichment facilities; insufficient opportunity and support for physical fitness; insufficient focus on retraining; employer and workplace attitudes
Quality of health service, e.g., waiting times	Change in delivery to meet demand; insufficient facilities for elder care; insufficient staffing for elder care

Two insights emerge from the right column of Table 9.1. The first is that several factors are repeated; they appear as drivers of several outcomes. The second is that it is possible to discern some structure. Table 9.2 shows one way of grouping the key factors. It places the factors from Table 9.1 into several general categories.

This analysis serves several purposes. It provides a schematic mapping of trends, determining factors, goals, and outcomes. It is an aid to both strategic assessment and strategic planning by creating a means for looking at the big picture while making clear the relationships among the main elements. Further, the groupings in Table 9.2 quickly enable one to distinguish areas that public policy might address in various forms (e.g., regulatory changes, priority budgeting of training activities) and those that government action would only indirectly affect or on which government would have little effect (e.g., demographics).

Robustness Assessment

Dynamic analyses of the type illustrated in the previous chapter and the scenario approaches discussed in this one bring home the value of explicitly considering the implications of different assumptions. This is crucial when moving toward the framing and assessment of policy alternatives, beyond the scope of this report. However, it is useful to touch briefly on this topic.

A robust plan is one that will meet specific expectations for positive outcomes across a range of possible futures. The property of robustness runs counter to traditional quantitative decision support as often practiced, which seeks optimal policies, ones likely to achieve the most favorable outcomes. The problem is that the set of assumptions that planners use to derive this optimum might not correspond to the conditions that could actually emerge during the course of the plan. Because of the complexities of the processes that yield socioeconomic outcomes, we cannot be certain that, if our original assumptions turn out to be only a little different from the actual future state of the world, our plan will work only a little less well than we had hoped. In fact, circumstances might be such that even a small difference could mean that

Table 9.2
Main Factors Driving Negative Outcomes, by Broad Category

Main Factor Cluster	Main Factor Driving Negative Outcome
Demography	Declining share of Israelis who are of working age
	Extended retirement years
	Insufficient labor participation
	Life-cycle patterns reducing consumption
	Specific pattern of aging in Israel
Society	Social patterns excluding elders
	Employer and workplace attitudes
	Reliance on families for elder support
	Too few models of work–life balance
Infrastructure	Insufficient facilities for elder care
	Insufficient support for physical fitness
	Insufficient elder-enrichment facilities
Budget	Increased costs of elder care
	Competing public social welfare needs
Regulation	Regulations on health care
	Regulations on employment
	Regulations on pension and retirement
	Change in delivery to meet demand
	Regulations on education
Planning and training	Insufficient focus on retraining
	Insufficient private retirement savings
	Lack of professional financial assistance
	Insufficient training in self-sufficiency
	Lack of opt-out–only saving programs
	Insufficient incentives for geriatric-worker training

the plan turns out to be completely ineffective or even worse. A robust plan, on the other hand, seeks satisfactory (versus optimal) outcomes for a greater chance of success.[2]

The search for robustness—the ability to achieve specific objectives under a wide range of circumstances—is well suited to the development of a strategic perspective and vice versa. Robustness comes closer to the way in which policymakers, as opposed to planners, view the world. They most often seek a hedged position that will sacrifice some degree of potential performance in exchange for greater certainty. In this sense, an analysis based on robustness might make an easier transition from the planners' analyses to the policy discussions of the senior political authorities. Working in the other direction, being mindful of the need for robustness might make the strategic perspective an important resource when developing actual government plans.

There are several means for seeking robustness in planning. An example of a nonquantitative technique is ABP (Dewar, 2002). ABP examines a strategy by identifying its underlying assumptions and then determines which of those are key in the sense of being so fundamental that, if they were to be violated, the entire strategy might risk failure. For example, a strategy that sought to increase the proportion of active workers in the labor force to the number of retirees would depend on assumptions about how well different incentives might affect, for example, individual retirement decisions and the extent to which employers would accept older workers.

From this, it becomes possible to inquire how vulnerable the key assumptions are (a technique used in the "what can go wrong" process we have described in this chapter). ABP then leads to a discussion of what additional actions might either provide a hedge against these prejudicial possibilities or, more actively, be used to shape external conditions so that the key assumptions have a greater likelihood of proving valid. A hedging strategy in the case of our strategy for broadening the base of workers might include preparations for novel alternative employment opportunities, such as creating new positions in the edu-

---

[2]   Although optimization is keyed to maximizing objectives, robustness employs the concept of satisficing—doing "well enough" (Simon, 1956).

cational system for older workers who can assist instruction in particular topics. A shaping strategy might be one that actively seeks to increase employers' willingness to accept older workers. In more-recent years, ABP methods have been used to generate ranges of alternative strategies rather than applying them only to test proposed strategies (see Lempert et al., 2008, for an example).[3]

Robust decisionmaking (RDM) is a method designed for use with quantitative models. It is not a new modeling technique but rather a different approach to the way models are used. Policymakers can combine factors that remain unknown or uncertain with the information they do have to support exploration of future conditions and the behavior of different strategies under those conditions with reference to the outcomes policymakers consider to be desirable.

The design of an RDM study and modeling system allows explicit statement of the factors to be considered: those representing external uncertainties over which the planners in the office conducting the study have no control,[4] the levers of policy or action that are under the control of the policymakers for whom the planners work, the relations between different factors that might themselves be unknown or about which different stakeholders might hold different views, and the measures that will be used to assess outcomes.

An RDM analysis of an aging Israel might seek to understand the relationships among the unknown factors, some of which are shown in Figure 9.4.

An analysis of this type was conducted to support examination and development of strategies for an aging population in the Netherlands (Auping et al., 2012; Logtens, Pruyt, and Gijsbers, 2012). Such outputs can make a large contribution to framing a strategic perspective, especially in a country that, like Israel, has already had direct experience with substantial demographic changes brought about by

---

[3] Another RAND project on strategies for use of natural gas in Israel (Popper et al., 2009) conducted an ABP workshop to kick off the project on its first day.

[4] This judgment is a relative one. Something that might be outside the jurisdiction of one ministry and so represent an external factor for planning purposes might be, for another, precisely its main purpose and so be classified as a lever at its disposal and discretion.

Figure 9.4
Simple Design for Robust-Decisionmaking Analysis of Population Aging in Israel

Major external factors	Levers for policy and action
Advances in life expectancy	Regulations on pensions and retirement
Growth of global and local economies	Investment in infrastructure
Labor-force participation rates	Improving efficiencies
Workforce productivity	Outreach to increase healthy behavior

Relationships	Measures
Effect of government policy on behaviors	Quality of life
	GDP per capita
	Solvency of public retirement funds
	Cost of health and long-term care

RAND *RR488-9.4*

immigration and might need to prepare for similar surprises in the future.

This chapter has shown how the what if–type of inquiry derived from scenario approaches might, once more, be placed within a larger framework that encompasses a full range of socioeconomic concerns. Chapter Ten brings the process of gaining a strategic perspective to the point at which it can support meaningful policy discussions.

Moving Toward Strategic Planning and Implementation

In this chapter, we encapsulate the main findings from the six steps as discussed in previous chapters and synthesize where there might be points of useful intervention for Israel's government to consider. We then shift to issues of process. We first discuss the next steps in developing a government strategy for population aging. We conclude by mapping the steps for doing so into the framework we recommended for strategic institutions and processes (Shatz et al., 2015) discussed in Chapter Two and that has now largely been put in place within the government of Israel. Whereas the focus of this report has been on the strategic perspective and thus the observe and orient steps of the OODA loop (as shown in Figure 2.2 in Chapter Two), the last portion of this final chapter provides a cursory overview of how portions of the decide and act steps of the OODA loop could operate within the system of processes and institutions we have recommended.

Strategic Leverage Points for Aging Strategies in Israel

We have addressed several of the initial questions posed in prior chapters. The main conclusion is that aging of Israel's population could affect socioeconomic outcomes enough to rise to a level of strategic concern. We next consider where strategic leverage points for influencing outcomes might be found and what alternative strategies are available to guide policy.

Several findings emerge that might form the basis for alternative strategic concepts for action.

- Disability, disease, and mortality in postretirement years appear to be considerably affected by nongenetic factors. In other words, characteristics associated with aging appear subject to some degree of modification by outside interventions. This, in turn, could add a measure of control over increases in both public and private costs associated with long-term care for the elderly.
- If so, the relevant groups for policy are not just youth, working age, and retired. Those in age categories beyond retirement should be treated as two separate groups: the retired who enjoy relatively good health and functional ability and the "oldest old." The point at which transition from retired to oldest old occurs is also amenable to policy interventions.[1]
- Issues raised by aging might lie less with the oldest old than with retirees who are increasing in both number and share of population and living longer, whose good health (which would delay costs to the health and long-term health systems) might be sustained by appropriate interventions, and whose contribution to society might be enhanced by policy changes.
- Mental, physical, and emotional well-being is affected by increasing engagement in work, leisure, and social activities (Rowe and Kahn, 1997).
- Meaningful engagement in post- or semiretirement years might require changes in attitudes toward aging, the nature of employment, education and training, balance between working and leisure time, and policies and programs to affect employers' perceptions.
- The phenomenon of having four living generations in a family is growing more prevalent. This raises questions of resource-sharing

---

[1] For example, studies have shown a causal relationship between early retirement and decrease in cognitive ability (Rohwedder and Willis, 2010). However, also see Australian Government Productivity Commission, 2005, for a cautionary note on whether the pattern of declining age-specific disability rates might be expected to result in meaningful reductions in public costs as the elder population increases in the future.

in terms of both short-term budgeting of limited incomes and longer-term intergenerational transfers. It also creates the possibility of facing a cross-generational struggle for resources or avoiding one by employing policy tools to create cross-generational solidarity and mutuality of interests.

- Some characteristics distinguish Israel from other countries facing similar demographic transitions:
  - Israel is entering a period of aging population with relatively high fertility rates for a developed country. This is offset by the relatively large size of the cohort entering retirement, a phenomenon associated with the large waves of immigration it experienced earlier.
  - Although the percentage share of older citizens as part of the total population will remain low by OECD standards, it will increase. The absolute number of retired and elderly will double in a relatively short time period and present challenges even with relatively favorable dependency ratios.
  - Israel's population is highly heterogeneous, with some communities (the Arab sector and the *haredim*) having little experience having a significant share of their populations being in the older generations.
  - Israel has several ministries that deal with aspects of aging and associated issues, including a dedicated Ministry for Senior Citizens. (What is not distinctive to Israel yet nonetheless true is the lack of a common database among them.)
  - The cohort of professionals associated with geriatric care is itself aging, with new entrants likely to be below the rate of replacement in the short term.
  - Israelis see families as the main resource for elder care. This has the potential to put an increasing burden on families in coming years but could also increase awareness of issues of quality of care and financial responsibility and so create a constituency for public action.

Israel could take any of a range of different actions in each of the issue areas touched by population aging. More generally, actions in any of these areas appear to fall under four main strategic approaches, including these:

- **budgetary approaches** for addressing potential funding gaps. This would consist of testing and addressing shortcomings in public safety nets, encouraging the accumulation of private savings to prepare for retirement, and government adjustments to taxes, benefits, and incentives to affect behavior and demand for services.
- **improving efficiencies** in cost and timing of care delivery. The government would seek improvements in data-gathering and analysis; reorganization and institutional change, for example, to achieve more integrative, collaborative, and multidisciplinary care; and greater use of long-distance and in-home care.
- **broadening the support base.** This approach seeks to broaden the base of support for an increasingly older population. Many countries are looking to immigration and family policies in order to increase the number of productive members of society. Israel has an opportunity to broaden the base by upgrading the skills of members of ethnic and religious communities that hitherto have been marginal workforce participants. This could also include changing work–life balance and opportunities to extend working years.
- **improving outcomes** for health and aging. This approach looks beyond efficiency of care delivery and toward enhancing care efficacy. Measures could include promoting evidence-based standards and best practices of care; seeking to enhance social interaction, inclusive participation, and support of elders; and influencing behavioral choices regarding physical activities and eating habits, among others.

For one example, the issue of ensuring sufficient hospital beds for elderly patients can be addressed in several ways. A *budgetary* approach would seek sources of funding for building the infrastructure that would meet anticipated demand. Connected with this might be efforts to *broaden the base* of financial support to make the budgetary choices less onerous. Israel could achieve greater *efficiencies* by reserving the beds that do exist for those cases that have no alternative to hospitalization while exploring and promoting hospitalization alternatives. These might include encouraging the use of in-home hospitalization or making better use of institutional alternatives to hospitals. Finally, a broader approach to a potential lack of hospital beds is to seek better health *outcomes*. This approach would explore policies and programs for improving elderly health by, for example, promoting health awareness and morbidity prevention (through primary care), conducting comprehensive geriatric assessments in the community, and providing home-oriented medical care and rehabilitation. Such efforts would have the potential to reduce the burden of hospitalizing the oldest old by extending the period of health and capability enjoyed by the able retired.

We gathered a range of potential actions and policies from international experience, existing documents on aging in Israel,[2] and interviews and discussions held in connection with this project. We have categorized them under the four broad approaches presented above and listed in Appendix C. We offer them not as recommendations but as examples of elements that might play a part in a strategy based on each of the four approaches or as a part of a composite strategy.

All measures will likely have some budgetary component, as well as usually some aspect that enhances efficiencies, broadens the base of support, or improves outcomes. These categories are helpful only to the extent that they aid in the process of supporting the reasoning that can be used to develop an overarching strategy. Further, these directions are by no means mutually exclusive. A strategy could well include different weightings of all four by taking different elements and actions.

It is worth pointing out that, for population aging, as with other socioeconomic issues, there are normative questions that might be illuminated but not resolved through analysis. Rather, political choices must be made either by default or through more-conscious balanc-

---

[2] We include, among others, Habib, 2012; NII, 2011; Braude, 2003; Yehezkel, 2007; Committee on the Long-Run Macroeconomic Effects of the Aging U.S. Population et al., 2012; Brodsky, Raznitsky, and Citron, undated; and Bank of Israel, 2008.

ing. For the socioeconomic consequences of population aging, some of these normative questions are as follows:

- What level of aging-related public costs is acceptable?
- What standards of care are acceptable?
- What is an appropriate mix of public and private effort and funding?
- What elements of quality of life should receive priority?
- How much economic participation should be expected from people in later life?
- What constitutes "successful" aging?

Process for Strategy

We now consider how the strategic perspective on aging—or on any other socioeconomic long-term issue in Israel—might move toward actual strategic plan development and implementation in the form of government policies and actions. We look specifically at the process that is currently emerging in Israel.

The main institutional locus for the longer-term strategic thinking process would be the new Socioeconomic Strategy Unit (Strategy Unit) within the NEC authorized in October 2012. In the context of population aging, the Strategy Unit would monitor various aging-related factors, such as population projections, health indicators, and labor-market indicators, on an ongoing basis. As part of its horizon scanning and foresight practice, it would identify trends and potential surprises.

The Strategy Unit would constantly work with a wide array of partners, both within the government and outside of it. Ministry-level strategy teams under deputy DGs would conduct similar processes on a narrower scale. In the context of aging, the main relevant ministries are the Ministry for Senior Citizens, Ministry of Health, Ministry of Social Affairs and Social Services, and the NII, but others might be involved based on the decision of the NEC director and the Professional-Level Strategy Forum (Professional Forum) made up of ministry

DGs that governs the Strategy Unit. The Bank of Israel, the Economics and Research Department of the Ministry of Finance, and the CBS would also support the process with data and research inputs.

In Shatz et al., 2015, we recommended an institution separate from government that would assist the Strategy Unit in gaining external engagement on this process. We suggested two options. The first would be a council of stakeholders, composed of representatives of civil society groups (business, labor, and third sector) and experts. It would discuss the issue, commission papers, hold hearings, or conduct other activities that would help members form their views. It would communicate its findings to the Strategy Unit and the Professional Forum. The second option would be an independent, smaller council of experts. It would investigate the issue and publish its findings, to which the Strategy Unit would respond. We refer to either format as an "external council."

The products of this process would be documented in the Strategy Unit's annual report. Papers and background research would be publicly released except in cases in which there were a compelling interest for keeping them internal to the government.

Informed by its ongoing work, overseen by the NEC director and the Professional Forum, if experts identify aging as an area of priority and desired action, the Strategy Unit, together with the relevant ministries and partners, would devise strategic alternatives for aging policy. The Strategy Unit would preliminarily assess the robustness of the alternatives and develop an understanding of potential shortcomings. This would probably require several iterations before a set of candidate robust alternatives emerges. This result would be presented, in coordination with the Professional Forum, to the PM-led Ministerial Strategy Forum (Political Forum) in its periodic meeting (and especially at the beginning of its term) to prioritize it within the government's agenda and action plans. Approval by the Political Forum would lead to further development of a detailed plan as part of the government's strategic agenda.

The strategic agenda would emerge from a **term-specific strategy planning** process designed to produce a term-specific plan, as coordinated by the Strategy Unit, assisted by its partners.

The Strategy Unit would take inputs from ministries, the external council, and the MOF Budget Department (among others) to write a first draft of this strategy, bracketing points of disagreement and estimating expected net costs. The Strategy Unit would present its first-draft staff paper to the Professional Forum and design an initial work plan, highlighting the key issues and problems.

The Professional Forum would approve or modify the key issues and problems and the initial work plan. This plan would then be included in the government's official work plan, which, in the beginning of the term, is called the socioeconomic agenda. The Professional Forum would begin also to formulate dedicated task forces (strategic plan teams) for selected issues of priority, which would manage and coordinate the detailed planning and execution of the strategic plan. These teams would be made up of PMO staff, external experts, and ministry experts, some of whom could be temporarily detailed to the PMO.

A strategic plan team would be the body leading the detailed formulation of the strategic plan in its selected area. It would do so with the close assistance of the Strategy Unit, the PMO staff, and external advisers, ministerial units, and especially MOF budget staff for determining scope and sources of revenue. The Professional Forum would iterate with strategic plan team staff to refine the strategic plan for presentation to the Political Forum.

At this point, the external institution discussed earlier (either the council of stakeholders or of experts, whichever has been created) could be asked to submit a review and evaluation of the plan. The Political Forum would be the ultimate authority reviewing and approving the detailed strategic plan and bringing it to the entire cabinet for final approval.

During the actual execution of the plan, initial management and coordination comes from the strategic plan team until responsibility can be transferred back to specific ministries or at the end of the term, whichever comes earlier. Subsequent governments would almost certainly revisit the strategic plan and could continue or discontinue execution, or revise the plan. However, they would need to publish justifications for their actions and revisions.

During execution, designated external evaluation teams, assisted by the Bank of Israel and the CBS, could monitor and evaluate outcomes. In parallel, the Strategy Unit would continue to conduct its long-term process of monitoring the effects that major factors are having on the society and economy and of performing foresight and horizon scanning. Some of these practices would be directly related to the newly established strategic plan.

Figure 10.1 provides a schematic and general timeline for the process we have outlined. It shows the roles played by the various bodies involved in strategic planning, as well as their degrees of involvement at different stages of the process. The first column shows the framework we have used in this report to describe the development of a strategic perspective and strategic plans, as reflected in the OODA process discussed in Chapter Two. The upper block represents the activities to maintain a strategic perspective that are ongoing and iterative. The lower block contains activities that are more related to specific acts of decision, planning, and implementation, as described in this section. For each step, we indicate which of these bodies takes the lead and which cooperate in some way with the lead institution. Obviously, this characterization will never fully capture the true interaction between activities or between different institutional bodies. Nevertheless, Figure 10.1 can serve as a guide to both the process and the distribution of effort in developing strategic perspectives and plans.

Embedding Strategy in Israel's Socioeconomic Policy

Like all other modern democracies, Israel must ensure that its governmental institutions and processes meet its citizens' needs while domestic and global environments change rapidly. In Israel, this challenge is perhaps greater than in most because of the relative newness of its governance apparatus, the rapidity of its development, and the circumstances of its geopolitical status. In the domestic realm, the achievement of continued socioeconomic progress could well require changes in approach reflective of the complexities and uncertainties with which Israel has been presented.

Figure 10.1
Roles of Government Bodies Involved in Strategic Planning

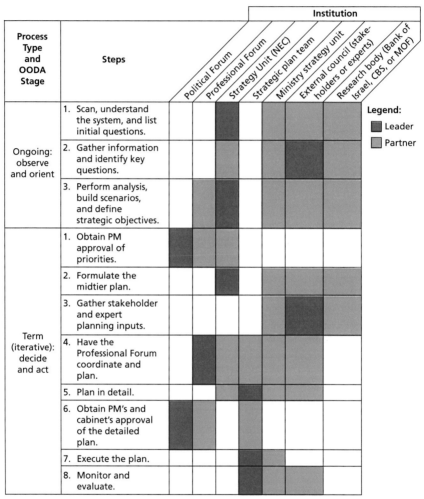

RAND RR488-10.1

In this report and its companion (Shatz et al., 2015), we lay out a framework of institutions and processes to meet the challenges of determining short-term policies and actions needed for a longer-term perspective that is rooted in objectives selected by elected policy leadership. This report has focused on presenting the type of effort that would characterize a long-term strategic thinking process. It is consistent with an ongoing process of active observation (especially for strategic early warning) and orientation through development of analyses and strategic alternatives.

The discussion we have presented is not intended as a blueprint to be followed in applying each detail. No such recipe will suffice for the task of guaranteeing a strategic perspective on all the socioeconomic issues government seeks to address. Indeed, the dynamics involved almost guarantee the ongoing presence of unresolved (and perhaps unresolvable) tensions. Rather, what we have sought is to motivate a style of approach, one that must be adapted to the particular circumstances in Israel as they evolve. The key concept is both the possibility and necessity of framing major policy and plans on an informed strategic perspective. All else follows from this course if it is pursued not as supplement to government process but as a necessary constituent element of that process.

The result that emerges from applying a strategic perspective to policy is not just a plan in the form of a document that can serve as the basis for operational planning and as a guideline for policy implementation. It will also support a process during which integration across ministries can occur and potential problems be revealed and worked out. It will provide the antecedent to developing explicit indicators for measuring progress, outcomes, and assessing whether and when modifications to plans and policies might be required.

Such an approach will not remove the inevitable policy controversies that can arise in such a heterogeneous society as that of Israel. But it will permit greater focus on the issues Israel confronts by creating a more widely shared basis for formulating and discussing alternative actions. This might not only provide more foundation for concerted public action but also allow the government to engage with the people of Israel in a manner that provides more opportunity for sharing insight and concerns and mobilizing consensus.

Candidate Indicators Within the Balanced Scorecard

Table A.1 provides a set of candidate indicators based on measures of socioeconomic outcomes in Israel across all four of the main accounts considered in Chapter Three. As before, this is intended to provide an illustration; it is by no means definitive, complete, or necessarily the only possible such framework. For example, we discussed weaknesses of each suggested metric and identified similar metrics used in OECD or other global indices, but we omit them here for ease of presentation. The intent is to provide a concrete example of the concepts discussed in Chapters Three and Four.

Table A.1
Candidate Indicators of Socioeconomic Outcomes for All Four Balanced-Scorecard Accounts

Goal	Dimension	Ideal Metric	ID	Suggested Measure
Enhancing socioeconomic well-being of individuals	Income and wealth	Availability of material means to achieve life goals	S-1	Household net adjusted disposable income per person
			S-2	Household discretionary income per person
		Degree of income inequality	S-3	Gini coefficient[a]
			S-4	Share of population below low-income threshold
			S-5	Average share of income shortfall below poverty threshold

Table A.1—Continued

Goal	Dimension	Ideal Metric	ID	Suggested Measure
Enhancing socioeconomic well-being of individuals, continued	Income and wealth, continued	Wealth to sustain shocks and meet goals over time	S-6	Household financial net worth per person
	Jobs	Availability of jobs commensurate with skills providing sufficient earnings to meet needs	S-7	Employment rate
			S-8	Long-term unemployment rate
			S-9	Involuntarily part-time work
		Quality of available jobs	S-10	Employees working on temporary contracts
			S-11	Average annual earnings per employee
	Housing	Access to adequate housing	S-12	Number of rooms per person
			S-13	Dwelling without basic facilities
		Housing satisfaction	S-14	Self-reported satisfaction with housing
		Housing affordability	S-15	Housing costs as share of adjusted disposable income
	Health status	Status with respect to disease	S-16	Life expectancy at birth
			S-17	Self-reported health status
		Status with respect to wellness	S-18	Self-reported limitations on activity
			S-19	Incidence of dementia and cognitive deterioration
			S-20	Rates of overweight and obesity

Table A.1—Continued

Goal	Dimension	Ideal Metric	ID	Suggested Measure
Enhancing socioeconomic well-being of individuals, continued	Work–life balance	Time balance between paid work, time with family, commuting, leisure, and personal care	S-21	Employees working very long hours
			S-22	Time devoted to leisure and personal care
			S-23	Average commuting time
			S-24	Employment rate of women with children of compulsory school age
	Education and skills	Educational attainment	S-25	Share of people between the ages of 25 and 64 with upper secondary degrees
			S-26	Share of people between the ages of 25 and 34 with tertiary education
		Students' cognitive skills	S-27	Programme for International Student Assessment reading skills
			S-28	Trends in International Mathematics and Science Study math and science skills
		Capacity for and availability of lifelong learning	S-29	Share of those 25 years and older who have engaged in further education in life skills, work skills, and general culture
		Civic learning	S-30	Independent testing of civic understanding and awareness
		Early-childhood development	S-31	Index of early-childhood development criteria

Table A.1—Continued

Goal	Dimension	Ideal Metric	ID	Suggested Measure
Enhancing socioeconomic well-being of individuals, continued	Education and skills, continued	Early-childhood development, continued	S-32	Class size in primary and lower secondary education
	Civic engagement and governance	Level of citizen engagement in the democratic process	S-33	Voter registration and voter turnout
			S-34	Voice and accountability survey
		Level of trust, transparency, and effectiveness of public policy	S-35	Consultation on rule-making
			S-36	Confidence in government, courts, and media
			S-37	Rule of law
	Social and cultural connections	Social network support	S-38	Availability of help through social contacts and families
		Frequency of social contact	S-39	Meet with social contacts and family at least once per week
		Social capital	S-40	Share who reply that most people can be trusted
	Environmental quality	Water and air quality	S-41	Microorganism and dissolved-metals assays
			S-42	PM_{10} concentration
		Environmental health impacts	S-43	Environmental burden of disease
		Parks and natural recreation areas	S-44	Access to recreational area
	Personal security	Experience of crime	S-45	Rates of intentional homicide
			S-46	Child deaths due to negligence, maltreatment, or physical assault

Table A.1—Continued

Goal	Dimension	Ideal Metric	ID	Suggested Measure
Enhancing socioeconomic well-being of individuals, continued	Personal security, continued	Fear of crime	S-47	Self-reported victimization
	Subjective well-being	People's overall views of their own lives	S-48	Life satisfaction
		People's present sense of satisfaction	S-49	Self-reported placement on affect-balance scale
Sustainable, growing, and innovative economy	Sustainable growth	Domestic economy	E-1	Real GDP growth per capita
			E-2	Gross fixed capital formation as a share of GDP
			E-3	Foreign direct investment
			E-4	Labor productivity
		Foreign trade	E-5	Real growth of exports of goods and services
			E-6	Current account balance
		Environmental sustainability	E-7	Production-based CO_2 emissions
			E-8	Demand-based CO_2 emissions
			E-9	Nitrogen surplus intensity
			E-10	Freshwater abstraction
	Innovation	Innovative goods and products	E-11	Shares of manufacturing output by level of technology
			E-12	Share of exports rated as high technology

Table A.1—Continued

Goal	Dimension	Ideal Metric	ID	Suggested Measure
Sustainable, growing, and innovative economy, continued	Innovation, continued	Innovative capacity growth	E-13	Gross domestic expenditure on research and development as a share of GDP
			E-14	Educational expenditure per student as a share of GDP per capita
	Macro stability and finance	Fiscal and monetary balances	E-15	Consumer price index growth
			E-16	Deficit as a share of GDP
			E-17	Public debt as a share of GDP
			E-18	Dependency ratio
			E-19	Share of primary civilian expenditures in total government spending
		International debt and payments	E-20	International reserves as a share of external short-term debt
	Regulation and competition	Regulatory burden on product markets	E-21	Assessment of state control over economy
			E-22	Assessment of barriers to entrepreneurship
			E-23	Assessment of barriers to trade and investment
		Ability to enact and enforce rules	E-24	Regulatory effectiveness
Democracy, effective government, and access to opportunity	Political participation	Engagement in political process, by region and community	D-1	Civic engagement and governance measures (S-33–S-37), by region and community

Table A.1—Continued

Goal	Dimension	Ideal Metric	ID	Suggested Measure
Democracy, effective government, and access to opportunity, continued	Government performance	People's perceptions of the adequacy of services	D-2	Waiting times for service, e.g., transfers from emergency room to hospital bed
			D-3	Governmental effectiveness
		Government efficiency	D-4	Government efficiency
		Citizens' perception of government corruption	D-5	Control of corruption
	Equal access to education	Effective access to information	D-6	Share of households with computers and access to the Internet, by region and community
		Effective access to high-quality education, by region and community	D-7	
	Equal access to employment	Effective access to jobs commensurate with skills, by gender	D-8	Gender gap in wages
		Effective access to jobs commensurate with skills, by region and community	D-9	
	Variations in quality of life	Regional and community variation in S-series (such as S-48–S-49) indicators	D-10	
Israel's role among the Jewish people	Partnership with Jewish communities	Involvement in Jewish people's projects	J-1	Share of government budget identified with Jewish existence and expression

Table A.1—Continued

Goal	Dimension	Ideal Metric	ID	Suggested Measure
Israel's role among the Jewish people, continued	Partnership with Jewish communities, continued	Involvement in Jewish people's projects, continued	J-2	Measures of burden-sharing with other Jewish communities
	Connection with Jewish individuals	Engagement with Israel in external Jewish communities	J-3	Annual visits to Israel, by country
		Israel's attractiveness to Jewish individuals	J-4	Immigration from high-income countries
	Jewish culture and education	Israel's leadership in sustaining and adding to Jewish cultural resources	J-5	

NOTE: CO_2 = carbon dioxide.

[a] The Gini coefficient is a statistical representation of income distribution and is often used to indicate inequality.

Selected Output from the November 2011 Backcasting Workshop

In this appendix, we report on some of the major themes that emerged from the backcasting workshop. In each case, we highlight responses that appeared common (in this case, mentioned by two of the working-group reports) among the working teams, as well as those that were distinctive to only one group.

"How Did the Article from the International Economics Newspaper Describe the 'New' Israel of 2027?"

Common Responses
- Israel's economic growth and GDP per capita
- closing gaps in income and opportunity
  - proportion of the poor in society and poverty reduction
  - integration of citizens at the bottom of society
- satisfaction of most citizens, as measured by indices of happiness and individual well-being, factoring in a variety of economic indicators
  - employment, including more inclusion of currently peripheral groups
  - cost of living (e.g., a few years to work to buy an apartment)
  - balanced mix of work and life ("life sane")
  - personal safety (e.g., crime and corruption)
  - health, culture, and environment
- high level and wide spread of educational attainment.

Less Common Responses
- world center in knowledge, computing, and technology (e.g., providing value to the world, beyond comparative to an absolute advantage)
- "However, there will remain the untreated areas, such as the field of aging with dignity and its psychological implications."
- reasonable national debt
- the image of Israel in the world
- increasingly attractive as a good place to live
  - "Jewish people's state of choice."

"What Occurred in the Intervening Years to Bring About This Israel of 2027?"

Common Responses
- better utilization and retention of human capital in economic activity
- expanding the education base—core curriculum and improving teacher status (not just human capital development more widely but also more focus on core knowledge)
- overcoming barriers of bureaucracy
  - changes in bureaucratic processes between public bodies and citizens and businesses (not only the removal of barriers but also the creation of options)
- promoting different industries that were able to flourish
- cultural change in the nature of social, political, and economic interaction.

Less Common Responses
- coordination between government departments, academia, the market, and the third sector
- significant change in the labor market ("improvement" on the "Danish" market model of flexibility and security, e.g., easier to fire and easier to recover from unemployment)
- infrastructure

- improving mobility—public transport
- effective urban construction to efficiently enhance the quality of life.

"What Underlying Factors Were Necessary for This Change to Occur?"

Common Responses
- created mutual trust between the strata of the population and between the population and government and sectors
- developed both solidarity and a sense of personal responsibility
- reconciliation with geopolitical environment—political process and international legitimacy
- changing priorities
  - less security and more education.

Unusual Responses
- nurtured the human capital of the elderly (utilizing increasing life expectancy to find ways for them to participate)
- personal example of leaders.

Candidate Policy Actions for Population Aging

Budgetary Approaches

Hospitals and Medical Training

- Increase the number of hospital beds and facilities intended for the elderly.
- Establish new geriatric departments and hospitals.
- Give significant incentives for internships and employment within the geriatric field.
- Increase the number of nurses at all levels of training.
- Create a mandatory module in geriatrics within nursing-school programs.
- Institute training tracks for nurses' aides, whose responsibilities would be parallel to those of practical nurses.
- Create a specialization track intended for nurses in geriatrics, who would function with maximum independence within the geriatric medical centers and in the community.

Community Care Services

- Expand existing elder-care services (e.g., supporting communities) with the intent of creating comprehensive sources of assistance and information encompassing the various needs of the nonhospitalized elderly population in an integrative manner.
- Develop 24-hour-response capabilities for the elderly in need.

- Expand and strengthen the health services in the community: Expand home care, and improve the care in all primary care services.
- Develop more rehabilitation centers within the community.

Informal Care

- Offer economic incentives for informal caregivers (e.g., compensation for loss of income and flexible employment conditions in the workplace).

Pensions and Labor Supply

- Through appropriate pensions, attractive working conditions, and possibly older-worker subsidies to employers, increase the labor-force participation rate among older workers and remove disincentives to work.
- Offer financial bonuses for late retirement and impose fines for early retirement.
- Set the retirement age in relation to changes in life expectancy.
- Enhance educational and skill-building opportunities in the Arab and *haredi* sectors to raise worker productivity and expand the base to support the costs of care for elderly.

Home Care

- Change income tests of eligibility for public support for home care services by using them solely to determine level of eligibility or dispensing with income tests and determining level of eligibility according to functional status.

Finance

- Reexamine the balance between public and private funding for financing elderly care, with particular attention to likely future trends.

Improving Efficiency

- Promote hospitalization alternatives, and encourage the use of home hospitalization.
- Construct a plan for training in geriatric medicine, and integrate it within internal, family, and emergency medical residencies.
- Establish a coherent professional role definition for all caregivers to the elderly. The creation and characterization of this professional field should include defining the following:
  - study and training
  - qualification
  - licensing
  - enforcement
  - advanced study and follow-up training
  - measures for preventing job burnout.
- Define an agreed process of geriatric assessment as a basic instrument for diagnosis and care plan development relating to the needs of the elderly and their families.
- Generate more-flexible and more-comprehensive services for elderly residents with disabilities (e.g., day centers).
- Develop a strategic plan for the management and supply of health and care information.

Broadening the Support Base

- Alter employers' approach to the older worker.
  - Promote an agenda for preventing age-based discrimination, advertise the advantages of employing elderly workers, and publish relevant studies and op-ed pieces in various media outlets.
  - Legislate against discrimination, and simultaneously legally permit reduced wages for the elderly.
- Improve employment conditions for elderly workers.
  - Develop training programs for specific professions, entrepreneurship, and facilitating second careers.

  – Conduct employment fairs for job seekers, and promote improved working conditions appropriate to the elderly population.
- Alter the common perception that the number of jobs in the economy is fixed (that is, that a job held by an older worker denies work to a younger one).
- Lower the cost of elderly employment.
- Adapt methods that allow employers to create more-flexible jobs in terms of working hours and the responsibility each job entails.
- Create a policy bank that will gather solutions for increasing elderly labor-force participation rate from countries throughout the world.
- Increase the labor-force participation rate within populations that have experienced low participation rates, e.g., the Arab female and Orthodox male populations.

Improving Outcomes

- Add to the educational specialization in internal and family medicine a mandatory training period of one month in a geriatric center, included within the residency year of each new physician, mandating that resident physicians in internal and family medicine choose extensions in geriatrics.
- Develop a policy regarding the necessary attention that should be given to the needs of the elderly's family members.
- Increase efforts to inform the public about caring for the elderly.
- Inform physicians about the variety of available community services.
- Improve elderly health at the community level by promoting health awareness and morbidity prevention, generating comprehensive geriatric assessments in the community, and providing home-oriented medical care and rehabilitation.
- Convey adequate instructions to the family members of the elderly, prior to a caregiver's placement, about the caregiver's responsibilities to provide proper supervision.

- Promote the development and application of advanced technologies of networking and computerization for monitoring and empowering of the elderly in their homes.
- View informal caregivers as themselves a population under health risk and treat as such through prevention, diagnosis, and treatment services specifically designated for that purpose.
- Identify existing opportunities for the instruction and support of family members, such as through visiting nurses.
- Enhance the supportive environments for the elderly living in the community (e.g., supportive communities and sheltered housing)
- Provide a broader range of community housing opportunities.

Bibliography

Alavinia, Seyed Mohammad, and Alex Burdorf, "Unemployment and Retirement and Ill-Health: A Cross-Sectional Analysis Across European Countries," *International Archives of Occupational and Environmental Health*, Vol. 82, No. 1, October 2008, pp. 39–45.

Andersen, Torben M., *Fremtidens velfærd: Globalisering, analyserapport og debatoplæg* [Future welfare: Globalization, analysis report and discussion paper], Copenhagen: Velfærdskommissionen, March 2005. As of July 16, 2015: http://pure.au.dk/portal/en/publications/fremtidens-velfaerd--globalisering-analyserapport-og-debatoplaeg(23dc6910-aab9-11da-bee9-02004c4f4f50).html

Andersen, Torben M., and Lars Haagen Pedersen, *Distribution and Labour Market Incentives in the Welfare State: Danish Experiences*, Institute for Evaluation of Labour Market and Education Policy, Working Paper 2008:10, May 17, 2008. As of May 4, 2015: http://www.ifau.se/en/Research/Publications/Working-papers/2008/Distribution-and-labour-market-incentives-in-the-welfare-state---Danish-experiences/

Angerman, William S., *Coming Full Circle with Boyd's OODA Loop Ideas: An Analysis of Innovation Diffusion and Evolution*, Wright-Patterson Air Force Base, Ohio: Air Force Institute of Technology, Air University, Department of the Air Force, AFIT/GIR/ENV/04M-01, March 2004. As of May 4, 2015: http://www.dtic.mil/dtic/tr/fulltext/u2/a425228.pdf

Auping, Willem, Erik Pruyt, Jan Kwakkel, Govert Gijsbers, and Michel Rademaker, *Aging: Uncertainties and Solutions*, The Hague: Centre for Strategic Studies and TNO, Report 2012-10, 2012. As of May 4, 2015: http://www.hcss.nl/reports/aging-uncertainties-and-solutions/109/

Australian Bureau of Statistics, "Relationships Between Domains of Progress," last updated April 29, 2009; referenced January 25, 2013. As of May 6, 2015: http://www.abs.gov.au/ausstats/abs@.nsf/Lookup/1383.0.55.001Feature%20Article22008%20(Edition%202)?opendocument&tabname=Summary&prodno=1383.0.55.001&issue=2008%20(Edition%202)&num=&view=

————, *Measures of Australia's Progress 2011: Summary Indicators*, October 6, 2011; last updated November 21, 2012; referenced January 24, 2013. As of May 4, 2015:
http://www.abs.gov.au/ausstats/abs@.nsf/Lookup/
by%20Subject/1370.0.55.001-2011-Main%20Features-Home%20page-1

————, "Future Directions in Measuring Australia's Progress," in *Measures of Australia's Progress 2010*, last updated November 13, 2013. As of May 4, 2015:
http://www.abs.gov.au/ausstats/abs@.nsf/Lookup/by%20Subject/
1370.0-2010-Chapter-Future%20directions%20%20%20287%29

Australian Early Development Census, undated home page. As of May 7, 2015:
https://www.aedc.gov.au/

Australian Government Productivity Commission, *Economic Implications of an Ageing Australia*, Canberra, Research Report, April 12, 2005. As of May 4, 2015:
http://www.pc.gov.au/inquiries/completed/ageing/report

Balfour, Arthur James, letter to Lord Rothschild, November 2, 1917. As of May 7, 2015:
http://avalon.law.yale.edu/20th_century/balfour.asp

Bank of Israel, *Economic Developments During the Last Months* [in Hebrew], No. 120, February 2008, pp. 26–29.

Banks, James, Richard Blundell, Antoine Bozio, and Carl Emmerson, "Disability, Health, and Retirement in the United Kingdom," in David A. Wise, ed., *Social Security Programs and Retirement Around the World: Historical Trends in Mortality and Health, Employment, and Disability Insurance Participation and Reforms*, Chicago, Ill.: University of Chicago Press, 2012, pp. 41–78.

Bell, Simon, Ken Eason, and Pia Frederiksen, eds., *A Synthesis of the Findings of the POINT Project*, European Commission, April 1, 2011. As of May 4, 2015:
http://static1.1.sqspcdn.com/static/f/602222/11716484/1302736662387/POINT_
synthesis_deliverable+15.pdf?token=oowZ3AYRZKQV1%2FFgZ5C7y0r5CtQ
%3D

Ben-David, Dan, ed., *State of the Nation Report: Society, Economy and Policy in Israel 2011–2012*, Jerusalem: Taub Center for Social Policy Studies in Israel, December 2012. As of May 4, 2015:
http://taubcenter.org.il/singer-series-2012/

Benjamin, Daniel J., Ori Heffetz, Miles S. Kimball, and Nichole Szembrot, "Beyond Happiness and Satisfaction: Toward Well-Being Indices Based on Stated Preference," *American Economic Review*, Vol. 104, No. 9, September 2014, pp. 2698–2735.

Bourgeois-Pichat, Jean, article title unknown, *Population*, Vol. 3, 1952.

————, article title unknown, *Population Bulletin of the United Nations*, Vol. 11, 1978.

Boyd, John R., "The Essence of Winning and Losing," briefing, Chet Richards and Chuck Spinney, eds., January 1996; posted August 2010; referenced April 22, 2013. As of May 4, 2015:
http://pogoarchives.org/m/dni/john_boyd_compendium/
essence_of_winning_losing.pdf

Braude, Jacob, *Age Structure and the Real Exchange Rate*, Jerusalem: Research Department, Bank of Israel Discussion Paper 2000.10, October 2000. As of May 4, 2015:
http://www.bankisrael.gov.il/deptdata/mehkar/papers/dp0010e.pdf

———, *The Impact of Demographics on Public Spending in the Long Run* [in Hebrew], Jerusalem: Research Department, Bank of Israel Discussion Paper 2003.04, June 2003.

Brodsky, Johnny, Shirley Raznitsky, and Danniella Citron, *Examination of Issues Faced by Family Members of the Elderly: Characteristics of Management, Burden, and Programs of Assistance and Support* [in Hebrew], Jerusalem: Myers–Joint Distribution Committee–Brookdale Institute, undated.

Canadian Index of Wellbeing, *How Are Canadians Really Doing? Highlights: Canadian Index of Wellbeing 1.0*, Waterloo, Ontario: Canadian Index of Wellbeing and University of Waterloo, 2011.

Cannuscio, Carolyn, Jason Block, and Ichiro Kawachi, "Social Capital and Successful Aging: The Role of Senior Housing," *Annals of Internal Medicine*, Vol. 139, No. 5, Part 2, September 2, 2003.

Central Bureau of Statistics, *Statistical Abstract of Israel*, Jerusalem: Central Bureau of Statistics, 2010.

Chernichovsky, Dov, and Eitan Regev, "Israel's Healthcare System," in Dan Ben-David, ed., *State of the Nation Report: Society, Economy and Policy in Israel 2011–2012*, Jerusalem: Taub Center for Social Policy Studies in Israel, December 2012, pp. 507–569.

Coale, Ansley J., article title unknown, *International Union for the Scientific Study of Population*, Vol. 4, 1981.

Coale, Ansley J., and Guang Guo, article on the use of new model life-tables at very low mortality in population projections, *Population Bulletin of the United Nations*, Vol. 30, 1990.

Cobb, R. W., and J. F. Coughlin, "Transportation Policy for an Aging Society: Keeping Older Americans on the Move," in *Transportation in an Aging Society: A Decade of Experience—Technical Papers and Reports from a Conference, November 7–9, 1999, Bethesda, Maryland*, 2004, pp. 275–289. As of May 4, 2015:
http://trid.trb.org/view.aspx?id=702085

Committee on the Long-Run Macroeconomic Effects of the Aging U.S. Population, Board on Mathematical Sciences and Their Applications, Division on Engineering and Physical Sciences, and Committee on Population, and Division of Behavioral and Social Sciences and Education, *Aging and the Macroeconomy: Long-Term Implications of an Older Population*, Washington, D.C.: National Academies Press, 2012. As of May 4, 2015:
http://www.nap.edu/catalog/13465/
aging-and-the-macroeconomy-long-term-implications-of-an-older

Committee for Planning National Geriatric Services, *Report of the Committee for Planning National Geriatric Services: 2010–2020, 2020–2030*, Hadassah Hospital, 2011.

Council of the European Union, "Establishing a General Framework for Equal Treatment in Employment and Occupation," Directive 2000/78/EC, November 27, 2000. As of May 8, 2015:
http://eur-lex.europa.eu/LexUriServ/
LexUriServ.do?uri=CELEX:32000L0078:en:HTML

Danish Government, *Progress, Innovation and Cohesion: Strategy for Denmark in the Global Economy—Summary*, May 2006a. As of May 4, 2015:
http://www.stm.dk/multimedia/
PROGRESS_INNOVATION_AND_COHESION.pdf

———, *Denmark's National Reform Programme: First Progress Report—Contribution to EU's Growth and Employment Strategy (The Lisbon Strategy)*, October 2006b. As of May 4, 2015:
http://ec.europa.eu/social/BlobServlet?docId=6097&langId=en

Demeny, Paul, "A Perspective on Long-Term Population Growth," *Population and Development Review*, Vol. 10, No. 1, March 1984, pp. 103–126.

Dewar, James A., *Assumption-Based Planning: A Tool for Reducing Avoidable Surprises*, Cambridge, UK: Cambridge University Press, 2002.

Drezner, Daniel W., "The Challenging Future of Strategic Planning," *Fletcher Forum of World Affairs*, Vol. 33, No. 1, Winter–Spring 2009, pp. 13–26.

Dubik, James M., "A National Strategic Learning Disability?" *Army*, September 2011, pp. 19–22. As of May 4, 2015:
http://www.ausa.org/publications/armymagazine/archive/2011/9/Documents/
FC_Dubik_0911.pdf

Dublin, Louis I., *The Facts of Life: From Birth to Death*, New York: Macmillan, 1951.

Dublin, Louis I., and Alfred J. Lotka, *Length of Life: A Study of the Life Table*, New York: Ronald Press, 1936.

Duell, Nicola, David Grubb, Shruti Singh, and Peter Tergeist, *Activation Policies in Japan*, Organisation for Economic Cooperation and Development Publishing, Social, Employment and Migration Working Paper 113, December 7, 2010. As of May 4, 2015:
http://www.oecd-ilibrary.org/social-issues-migration-health/
activation-policies-in-japan_5km35m63qqvc-en?crawler=true

Eisenhower, Dwight D., *Public Papers of the Presidents of the United States: Dwight D. Eisenhower, 1957—Containing the Public Messages, Speeches, and Statements of the President, January 1 to December 31, 1957,* Washington, D.C.: Office of the Federal Register, National Archives and Records Service, General Services Administration, 1958. As of May 4, 2015:
http://quod.lib.umich.edu/p/ppotpus/4728417.1957.001?view=toc

European Foundation for the Improvement of Living and Working Conditions, "European Quality of Life Surveys (EQLS)," November 16, 2012.

European Union, "European Union Statistics on Income and Living Conditions (EU-SILC)," undated.

"Excerpt from an Address Before a Joint Session of Congress, 25 May 1961," Columbia Broadcasting System, May 25, 1961. As of May 4, 2015:
http://www.jfklibrary.org/Asset-Viewer/xzw1gaeeTES6khED14P1Iw.aspx

Finnish Consultative Committee on Road Safety, *Road Safety 2006–2010,* Helsinki: Ministry of Transport and Communications Finland, Programs and Strategies 1/2006, 2006. As of May 4, 2015:
http://www.lvm.fi/fileserver/road%20safety%202006-2010.pdf

Finnish Parliament, Universities Act, 558/2009, c. 2009. As of May 8, 2015:
http://www.finlex.fi/en/laki/kaannokset/2009/en20090558.pdf

Frejka, Tomas, *The Future of Population Growth: Alternative Paths to Equilibrium,* New York: John Wiley, 1973.

French Commission on the Measurement of Economic Performance and Social Progress, *Report of the Commission on the Measurement of Economic Performance and Social Progress,* September 14, 2009.

Friedberg, Aaron L., "Strengthening U.S. Strategic Planning," *Washington Quarterly,* Vol. 31, No. 1, Winter 2007–2008, pp. 47–60.

Fries, James F., "Aging, Natural Death, and the Compression of Morbidity," *New England Journal of Medicine,* Vol. 303, No. 3, July 17, 1980, pp. 130–135.

———, "The Compression of Morbidity: Near or Far?" *Milbank Quarterly,* Vol. 67, No. 2, 1989, pp. 208–232.

Gaddis, John Lewis, "What Is Grand Strategy?" prepared as the Karl Von Der Heyden Distinguished Lecture, Duke University, February 26, 2009.

Gallup, "Gallup World Poll," undated. As of May 7, 2015:
http://www.gallup.com/services/170945/world-poll.aspx

Glenn, Jerome C., "The Futures Wheel," in Jerome C. Glenn and Theodore J. Gordon, eds., *Futures Research Methodology Version 3.0*, Washington, D.C.: Millennium Project, 2009. As of May 4, 2015:
http://www.millennium-project.org/millennium/FRM-V3.html

Glenn, Jerome C., and Theodore J. Gordon, eds., *Futures Research Methodology Version 3.0*, Washington, D.C.: Millennium Project, 2009. As of May 4, 2015:
http://www.millennium-project.org/millennium/FRM-V3.html

Gruber, Jonathan, Kevin Milligan, and David A. Wise, *Social Security Programs and Retirement Around the World: The Relationship to Youth Unemployment, Introduction and Summary*, Cambridge, Mass.: National Bureau of Economic Research, Working Paper 14647, January 2009. As of May 4, 2015:
http://www.nber.org/papers/w14647

Habib, Jack, *12th Dead Sea Conference: National Preparedness for an Aging Population* [in Hebrew], Jerusalem: National Institute for Health Policy Research, 2012.

Hagist, Christian, and Laurence Kotlikoff, *Who's Going Broke? Comparing Healthcare Costs in Ten OECD Countries*, Cambridge, Mass.: National Bureau of Economic Research, Working Paper 11833, December 2005. As of May 4, 2015:
http://www.nber.org/papers/w11833

Hall, Stephen, Edward Tredger, Mohammad Ali, and Pat Thomas, eds., *Sustainable Development Indicators in Your Pocket 2009: An Update of the UK Government Strategy Indicators*, London: Department for Environment, Food and Rural Affairs, PB13265, 2009. As of May 4, 2015:
https://www.gov.uk/government/statistics/
sustainable-development-indicators-in-your-pocket-2009

Herzl, Theodor, *Altneuland* [Old new land], Leipzig: Hermann Seemann Nachfolger, 1902.

Hoorens, Stijn, Federico Gallo, Jonathan A. Cave, and Jonathan C. Grant, "Can Assisted Reproductive Technologies Help to Offset Population Ageing? An Assessment of the Demographic and Economic Impact of ART in Denmark and UK," *Human Reproduction*, Vol. 22, No. 9, September 2007, pp. 2471–2475.

International Civic and Citizenship Education Study, undated home page. As of May 7, 2015:
http://iccs.acer.edu.au/

International Institute for Democracy and Electoral Assistance, undated home page. As of May 7, 2015:
http://www.idea.int/

Irish Central Statistics Office, *Measuring Ireland's Progress 2010*, Prn A11/1564, September 2011. As of May 4, 2015:
http://www.cso.ie/en/releasesandpublications/othercsopublications/
measuringirelandsprogress2010/

Ishi, Kotaro, "Macrosocial Challenges in Israel," in International Monetary Fund, *Israel: Selected Issues Paper*, Washington, D.C.: International Monetary Fund, Country Report 12/71, April 2012, pp. 2–18. As of May 4, 2015:
https://www.imf.org/external/pubs/ft/scr/2012/cr1271.pdf

Jewish People's Council, "The Declaration of the Establishment of the State of Israel," May 14, 1948. As of May 7, 2015:
http://www.mfa.gov.il/mfa/foreignpolicy/peace/guide/pages/
declaration%20of%20establishment%20of%20state%20of%20israel.aspx

Jørgensen, Jan, and Henry Mintzberg, "Emergent Strategy for Public Policy," *Canadian Public Administration*, Vol. 30, No. 2, June 1987, pp. 214–229.

Kaplan, Robert S., and David P. Norton, "The Balanced Scorecard: Measures That Drive Performance," *Harvard Business Review*, January–February 1992, pp. 71–80.

———, *Balanced Scorecard: Translating Strategy into Action*, Cambridge, Mass.: Harvard Business School Press, 1996.

Kaufmann, Daniel, Aart Kraay, and Massimo Mastruzzi, *Governance Matters VIII: Aggregate and Individual Governance Indicators 1996–2008*, Vol. 1, Policy Research Working Paper WPS 4978, June 1, 2009.

Lafortune, Gaetan, and Gaëlle Balestat, *Trends in Severe Disability Among Elderly People: Assessing the Evidence in 12 OECD Countries and the Future Implications*, Paris: OECD Publishing, Organisation for Economic Cooperation and Development Health Working Paper 26, March 30, 2007. As of May 4, 2015:
http://www.oecd-ilibrary.org/social-issues-migration-health/
trends-in-severe-disability-among-elderly-people_217072070078

Lee, Ronald, and Andrew Mason, "Fertility, Human Capital, and Economic Growth over the Demographic Transition," *European Journal of Population*, Vol. 26, No. 2, May 2010, pp. 159–182.

Lempert, Robert J., Steven W. Popper, and Steven C. Bankes, *Shaping the Next One Hundred Years: New Methods for Quantitative, Long-Term Policy Analysis*, Santa Monica, Calif.: RAND Corporation, MR-1626-RPC, 2003. As of July 16, 2015:
http://www.rand.org/pubs/monograph_reports/MR1626.html

Lempert, Robert J., Horacio R. Trujillo, David Aaron, James A. Dewar, Sandra H. Berry, and Steven W. Popper, *Comparing Alternative U.S. Counterterrorism Strategies: Can Assumption-Based Planning Help Elevate the Debate?* Santa Monica, Calif.: RAND Corporation, DB-548-RC, 2008. As of May 4, 2015:
http://www.rand.org/pubs/documented_briefings/DB548.html

Logtens, Tom, Erik Pruyt, and Govert W. Gijsbers, "Societal Aging in the Netherlands: Exploratory System Dynamics Modeling and Analysis," in *Conference Proceedings: The 30th International Conference of the System Dynamics Society*, St. Gallen, Switzerland, July 10, 2012. As of May 4, 2015: http://www.systemdynamics.org/conferences/2012/proceed/papers/P1267.pdf

Luo, Ye, Louise C. Hawkley, Linda J. Waite, and John T. Cacioppo, "Loneliness, Health, and Mortality in Old Age: A National Longitudinal Study," *Social Science and Medicine*, Vol. 74, No. 6, March 2012, pp. 907–914.

Martinez, I. L., K. Frick, T. A. Glass, M. Carlson, E. Tanner, M. Ricks, and L. P. Fried, "Engaging Older Adults in High Impact Volunteering That Enhances Health: Recruitment and Retention in the Experience Corps Baltimore," *Journal of Urban Health*, Vol. 83, No. 5, September 2006, pp. 941–953.

McDougall, Walter A., "Can the United States Do Grand Strategy?" *Orbis*, Vol. 54, No. 2, 2010, pp. 165–184.

Ministry for Senior Citizens, *Economic Situation and Emotional Health of Persons Aged 65 and over in Israel, Compared to European Countries Based on the SHARE Survey of Health, Aging and Retirement Europe and Israel (Selected Findings)* [in Hebrew], undated; referenced June 12, 2013. As of May 4, 2015: http://www.vatikim.gov.il/data/Documents/SHARE%202011.pdf

Mintzberg, Henry, *The Rise and Fall of Strategic Planning: Reconceiving Roles for Planning, Plans, Planners*, New York: Free Press, 1994.

Montez, Jennifer Karas, Robert A. Hummer, and Mark D. Hayward, "Educational Attainment and Adult Mortality in the United States: A Systematic Analysis of Functional Form," *Demography*, Vol. 49, No. 1, February 2012, pp. 315–336.

National Insurance Institute, *Model for Examining the Financial and Social Stability of the National Insurance* [in Hebrew], Jerusalem, June 18, 2011.

NII—*See* National Insurance Institute.

Nordic Council and the Nordic Council of Ministers, "Danish Welfare Proposals," Copenhagen, December 8, 2005. As of May 4, 2015: http://www.norden.org/en/news-and-events/news/danish-welfare-proposals

OECD—*See* Organisation for Economic Co-operation and Development.

Oeppen, Jim, and James W. Vaupel, "Broken Limits to Life Expectancy," *Science*, Vol. 296, No. 5570, May 10, 2002, pp. 1029–1031.

Offord Centre for Child Studies, "The Early Development Instrument," undated. As of May 7, 2015: http://www.offordcentre.com/readiness/EDI_viewonly.html

Olshansky, S. Jay, Bruce A. Carnes, and Aline Désesquelles, "Predicting Human Longevity," *Science*, Vol. 292, No. 5522, June 1, 2001, pp. 1654–1655.

Organisation for Economic Co-operation and Development, "How's Life? Measuring Well-Being," undated (a). As of July 16, 2015:
http://www.oecd-ilibrary.org/economics/how-s-life_23089679

———, "Indicators of Product Market Regulation Homepage," undated (b). As of May 11, 2015:
http://www.oecd.org/economy/growth/
indicatorsofproductmarketregulationhomepage.htm

———, "Labour Statistics," undated (c). As of July 16, 2015:
http://www.oecd.org/employment/labour-stats/

———, *Ageing and Transport: Mobility Needs and Safety Issues*, Paris, October 25, 2001a. As of May 4, 2015:
http://www.oecd-ilibrary.org/transport/ageing-and-transport_9789264195851-en

———, *The DAC Guidelines: Guidance for Development Co-Operation—Strategies for Sustainable Development*, Paris, November 2001b. As of May 4, 2015:
http://www.oecd.org/dac/environment-development/
strategiesforsustainabledevelopment.htm

———, *Pensions at a Glance: Public Policies Across OECD Countries*, Paris, May 2, 2005. As of May 4, 2015:
http://www.oecd-ilibrary.org/finance-and-investment/
oecd-pensions-at-a-glance-2005_pension_glance-2005-en

———, *Live Longer, Work Longer*, Paris, 2006. As of May 4, 2015:
http://www.oecd-ilibrary.org/employment/
live-longer-work-longer_9789264035881-en

———, *Pensions at a Glance 2011: Retirement-Income Systems in OECD and G20 Countries*, Paris, March 17, 2011. As of May 4, 2015:
http://www.oecd-ilibrary.org/finance-and-investment/
pensions-at-a-glance-2011_pension_glance-2011-en

Oxley, Howard, *Policies for Healthy Ageing: An Overview*, Paris: Organisation for Economic Co-operation and Development Publishing, Organisation for Economic Co-operation and Development Health Working Paper 42, February 16, 2009. As of May 4, 2015:
http://www.oecd-ilibrary.org/social-issues-migration-health/
policies-for-healthy-ageing_226757488706

Panel on a Research Agenda and New Data for an Aging World, Committee on Population and Committee on National Statistics, Division of Behavioral and Social Sciences and Education, National Research Council, *Preparing for an Aging World: The Case for Cross-National Research*, Washington, D.C.: National Academy Press, 2001. As of May 4, 2015:
http://www.nap.edu/catalog/10120/
preparing-for-an-aging-world-the-case-for-cross-national

Parker-Pope, Tara, "For a Healthy Retirement, Keep Working," *Well*, October 19, 2009. As of May 8, 2015:
http://well.blogs.nytimes.com/2009/10/19/
for-a-healthy-retirement-keep-working/?_r=0

Pirttilä, Jukka, and Håkan Selin, *Tax Policy and Employment: How Does the Swedish System Fare*, Helsinki: Labour Institute for Economic Research Discussion Paper 267, 2011. As of May 4, 2015:
http://www.labour.fi/tutkimusjulkaisut/tyopaperit/sel267.pdf

PMO—*See* Prime Minister's Office.

Popper, Steven W., James Griffin, Claude Berrebi, Thomas Light, and Endy M. Daehner, *Natural Gas and Israel's Energy Future: A Strategic Analysis Under Conditions of Deep Uncertainty*, Santa Monica, Calif.: RAND Corporation, TR-747-YSNFF, December 2009. As of May 4, 2015:
http://www.rand.org/pubs/technical_reports/TR747.html

Prime Minister's Office, Policy Planning Division, *The Government Planning Guide, Version 4.1* [in Hebrew], Jerusalem, September 2010.

Rechel, Bernd, Yvonne Doyle, Emily Grundy, and Martin McKee, *How Can Health Systems Respond to Population Ageing?* European Observatory on Health Systems and Policies, Policy Brief 10, 2009. As of May 4, 2015:
http://www.euro.who.int/__data/assets/pdf_file/0004/64966/E92560.pdf

Rohwedder, Susann, and Robert J. Willis, "Mental Retirement," *Journal of Economic Perspectives*, Vol. 24, No. 1, Winter 2010, pp. 119–138.

Rowe, John W., and Robert L. Kahn, "Successful Aging," *Gerontologist*, Vol. 37, No. 4, 1997, pp. 433–440.

Rumelt, Richard P., *Good Strategy, Bad Strategy: The Difference and Why It Matters*, New York: Crown Business, 2011.

Sayle, Timothy Andrews, "Defining and Teaching Grand Strategy," *Telegram*, Vol. 4, January 2011. As of May 4, 2015:
http://www.fpri.org/articles/2011/01/defining-and-teaching-grand-strategy

Schnalzenberger, Mario, and Rudolf Winter-Ebmer, "Layoff Tax and Employment of the Elderly," Johannes Kepler University of Linz, Department of Economics, Working Paper 0819, November 2008. As of May 4, 2015:
https://ideas.repec.org/p/jku/econwp/2008_19.html

Schwartz, Peter, *The Art of the Long View: Planning for the Future in an Uncertain World*, New York: Currency Doubleday, 1991.

Secretary-General of the League of Nations, "Mandate for Palestine," London, July 24, 1922. As of May 7, 2015:
http://unispal.un.org/UNISPAL.NSF/0/
2FCA2C68106F11AB05256BCF007BF3CB

Shalev, Michael, Johnny Gal, and Sagit Azary-Viesel, "The Cost of Social Welfare: Israel in Comparative Perspective," in Dan Ben-David, ed., *State of the Nation Report: Society, Economy and Policy in Israel 2011–2012*, Jerusalem: Taub Center for Social Policy Studies in Israel, December 2012, pp. 383–394.

Shatz, Howard J., Steven W. Popper, Sami Friedrich, Shmuel Abramzon, Anat Brodsky, Roni Harel, and Ofir Cohen, *Developing Long-Term Socioeconomic Strategy in Israel: Institutions, Processes, and Supporting Information*, Santa Monica, Calif.: RAND Corporation, RR-275-PMO, 2015.

Siegel, Jacob S., paper title unknown, in Suzanne G. Haynes and Manning Feinleib, eds., *Second Conference on the Epidemiology of Aging: Proceedings of the Second Conference, March 28–29, 1977*, Bethesda, Md.: U.S. Department of Health and Human Services; Public Health Services; National Institutes of Health; National Institute on Aging; and National Heart, Lung, and Blood Institute, 1980.

Simon, Herbert A., "Rational Choice and the Structure of the Environment," *Psychological Review*, Vol. 63, No. 2, 1956, pp. 129–138.

Singapore Prime Minister's Office, Public Service Division, *Conversations for the Future*, Vol. 1: *Singapore's Experiences with Strategic Planning (1988–2011)*, Singapore, 2011.

Springate, Iain, Mary Atkinson, and Kerry Martin, *Intergenerational Practice: A Review of the Literature*, Slough, Berkshire, UK: National Foundation for Educational Research, Research Report, May 2008.

Steptoe, Andrew, Aparna Shankar, Panayotes Demakakos, and Jane Wardle, "Social Isolation, Loneliness, and All-Cause Mortality in Older Men and Women," *Proceedings of the National Academy of Sciences of the United States of America*, Vol. 110, No. 15, March 25, 2013, pp. 5797–5801. As of May 4, 2015: http://www.pnas.org/content/110/15/5797

Stessman, Yohanan, *Committee for Planning National Geriatric Arrangements, 2010–2020 and 2020–2030* [in Hebrew], Jerusalem: Ministry of Health, October 2011.

Stoll, Laura, Juliet Michaelson, and Charles Seaford, *Well-Being Evidence for Policy*, London: New Economics Foundation, April 3, 2012. As of May 4, 2015: http://www.neweconomics.org/publications/entry/well-being-evidence-for-policy-a-review

Struyk, Raymond J., and Harold M. Katsura, *Aging at Home: How the Elderly Adjust Their Housing Without Moving*, New York: Haworth Press, 1988.

Survey of Health, Ageing and Retirement in Europe; Hebrew University of Jerusalem; and Israel Gerontological Data Center, "הלועש יפכ הפוריא תונידמל האוושהב הפוריאב הלעמו 65 ינב לש תישגרו תיתואירב ,תילכלכ בצמ תנומת השירפהו הנקיזה ,תואירבה רקסמ לארשיבו)SHARE(- ממאצאימ נברחביברישאל, [Economic portrait and emotional health of those aged 65 and over in Israel compared to European evidence from the Survey of Health, Aging and Retirement (SHARE) (selected findings)]" [in Hebrew], undated. As of May 7, 2015:
http://vatikim.gov.il/data/Documents/SHARE%202011.pdf

Swiss Federal Statistical Office, *Sustainable Development Report 2012*, Neuchâtel, Report 1248-1200, December 2012. As of May 4, 2015:
http://www.bfs.admin.ch/bfs/portal/en/index/themen/21/22/publ.html?publicationID=4066

UK Department of Health, *Business Plan 2012–2015*, May 31, 2012.

UK House of Commons Public Administration Select Committee, *Who Does UK National Strategy? First Report of Session 2010–11: Report, Together with Formal Minutes, Oral and Written Evidence*, London: Stationery Office Limited, HC 435, October 18, 2010. As of May 4, 2015:
http://www.publications.parliament.uk/pa/cm201011/cmselect/cmpubadm/435/435.pdf

UK Prime Minister's Strategy Unit, *Strategy Survival Guide*, Version 2.1, Cabinet Office, July 2004. As of May 4, 2015:
http://webarchive.nationalarchives.gov.uk/20060213205515/http:/strategy.gov.uk/downloads/survivalguide/downloads/ssg_v2.1.pdf

United Nations, *World Population Prospects*, 1973.

———, *World Population Prospects*, 1986.

———, *World Population Prospects*, 1989.

———, *World Population Prospects*, 1999.

———, *World Population Prospects*, 2001.

———, *World Population Prospects*, 2008.

United Nations General Assembly, "Future Government of Palestine," A/RES/181(II), November 29, 1947. As of May 7, 2015:
http://unispal.un.org/unispal.nsf/0/7F0AF2BD897689B785256C330061D253

United Nations Office on Drugs and Crime, "United Nations Surveys on Crime Trends and the Operations of Criminal Justice Systems (CTS)," undated. As of May 7, 2015:
http://www.unodc.org/unodc/en/data-and-analysis/United-Nations-Surveys-on-Crime-Trends-and-the-Operations-of-Criminal-Justice-Systems.html